The Future of Animal Farming

W9-CBB-969

DATE DUE

This book is dedicated to Ruth Harrison (1920–2000)

and David Wood-Gush (1922–1992)

"Concern for farm animal welfare is a virtue, but often seen to be in conflict with other virtues; environmental sustainability and economic justice. The value of this book is that it sees all three as facets of the same thing: respect for life. From this firm base it offers a moral and a sensible guide to the future. It explains not only why we should all care for all life on the land but also how we may all contribute to that care."

John Webster, Emeritus Professor, University of Bristol

"This book should be read by anyone interested in animal welfare, and in the environment. 'Factory farming' is rightly receiving much criticism but what are the ethics of rearing animals in this new century, and what are the practical answers to treating our animals properly? With considerable editorial skill Marian Dawkins and Roland Bonney have drawn together articles encapsulating the views of campaigning groups, of farmers and of retailers. What emerges is a consensus for soundly based farming methods which are humane. The really interesting conclusion is that such methods bring economic advantages to those who introduce them."

Sir Brian Follett, University of Oxford

The Future of Animal Farming

Renewing the Ancient Contract

Edited by

Marian Stamp Dawkins
Somerville College and Department of Zoology
University of Oxford
Oxford, UK

Roland Bonney
Food Animal Initiative
The Field Station
Wytham, Oxon, UK

Blackwell
Publishing

BLACKWELL PUBLISHING
350 Main Street, Malden, MA 02148-5020, USA
9600 Garsington Road, Oxford OX4 2DQ, UK
550 Swanston Street, Carlton, Victoria 3053, Australia

The right of Marian Stamp Dawkins and Roland Bonney to be identified as the authors of the editorial material in this work has been asserted in accordance with the UK Copyright, Designs, and Patents Act 1988.

Designations used by companies to distinguish their products are often claimed as trademarks. All brand names and product names used in this book are trade names, service marks, trademarks, or registered trademarks of their respective owners. The publisher is not associated with any product or vendor mentioned in this book.

This publication is designed to provide accurate and authoritative information in regard to the subject matter covered. It is sold on the understanding that the publisher is not engaged in rendering professional services. If professional advice or other expert assistance is required, the services of a competent professional should be sought.

First published 2008 by Blackwell Publishing Ltd

1   2008

Library of Congress Cataloging-in-Publication Data

The future of animal farming : renewing the ancient contract / edited by Marian Stamp Dawkins and Roland Bonney.
    p. cm.
    Includes bibliographical references and index.
    ISBN 978-1-4051-8583-7 – ISBN 978-1-4051-7782-5 (pbk. : alk. paper) 1. Livestock. 2. Animal welfare. I. Dawkins, Marian Stamp. II. Bonney, Roland.

    HV4757.F88   2008
    179.3–dc22

                                    2007040797

A catalogue record for this title is available from the British Library.

Set in 10/13pt RotisSemiSans
by SPi Publisher Services, Pondicherry, India
Printed and bound in Singapore
by COS Printers Pte Ltd

The publisher's policy is to use permanent paper from mills that operate a sustainable forestry policy, and which has been manufactured from pulp processed using acid-free and elementary chlorine-free practices. Furthermore, the publisher ensures that the text paper and cover board used have met acceptable environmental accreditation standards.

For further information on
Blackwell Publishing, visit our website at
www.blackwellpublishing.com

Contents

Foreword

Those who know my views from *Animal Liberation* may be surprised to find me writing a foreword to a book entitled *The Future of Animal Farming.* Doesn't the animal liberation movement do its very best to ensure that animal farming has no future? If the correct moral principle for guiding our conduct towards nonhuman animals is to give their interests equal consideration with our own, at least where our interests are similar, should we be farming them at all? Shouldn't we all be vegans?

Then there is the book's subtitle: *Renewing the Ancient Contract.* How could there really be a contract between humans and animals? The idea of a contract presupposes that both parties choose to enter it. Cows, pigs, and chickens don't have the capacity to make an informed choice about whether or not to associate with humans – to mention just one critical fact, they cannot know that their premature death is part of the bargain. The reality is that domestic animals have always been captured, bred, reared, and killed for the benefit of humans, and rarely, if ever, given the opportunity to break free and live on their own. The slave trade is a closer parallel to this than the modern idea of an agreement between freely contracting parties. No doubt traditional farmers were more likely to care for their animals as individuals than the people who manage today's vast factory farms, but the fact that some slave-owners had a genuine, if paternalistic, concern for the welfare of their slaves was not enough to make the slave trade a contract between Europeans and Africans.

I do not resile from the position I took in *Animal Liberation.* I see the rearing of animals for food as a manifestation of speciesism, that is, a human prejudice against giving proper consideration to the interests of beings of other species. Commercial animal raising is inherently likely to sacrifice the interests of the animals to our own convenience. Refusing to buy animal products is the surest way to avoid supporting the unethical treatment of animals. But I also recognize that while the number of vegetarians and vegans has grown, at least in developed countries, during the past three decades, the number of animals raised and killed for food, worldwide, has grown even faster. This is in large part because of greater prosperity, both in the developed world and in countries like China, and the higher demand for animal products that this prosperity brings. That demand has in turn led to a staggering increase in the number of animals spending miserable lives in the close confinement of factory farms.

In the face of this vast universe of animal suffering – which is also an ecological catastrophe on many different levels, from local water pollution to the acceleration of climate change – should the animal movement confine itself to promoting veganism? Over the next 10 or 20 years that strategy may, with luck, increase the percentage of vegans to 5% or even 10% of the population, but on the basis of what we have seen so far, the chances of it succeeding in persuading the majority of meat-eaters to abandon all animal products are remote. (At least, unless the development of *in vitro* meat offers a more economical but otherwise indistinguishable alternative to meat derived from animals.) This means that during the next decade or two, billions of animals will live and die in factory farms, their numbers barely diminished by the slowly growing number of vegans, and their sufferings entirely unaffected by it.

It therefore seems better to pursue a different strategy. We should do our utmost to reduce the suffering of those billions of animals. This is not an either/or choice. The animal movement should continue to promote a cruelty-free vegan lifestyle, and to encourage those who are not vegans to eat less meat and dairy products. Recognizing that not everyone is ready to make such changes, however, the movement should also be involved in improving the welfare of animals used in commercial farming.

This strategy can succeed. While I was writing this foreword Oregon became the third state in the US to ban sow stalls – known in America as gestation crates – which are commonly used to confine pregnant sows in metal crates too small for them even to turn around or walk a few steps. Earlier, Florida and Arizona had passed similar bans as a result of referenda initiated by the signatures of large numbers of voters. Significantly, the law in Oregon was the first in the US to come about through the normal process of representative democracy at the state level. The European Union and Australia have also agreed to prohibit sow stalls for most of the sow's pregnancy. In addition to these legal changes, the suffering of an even larger number of pigs will in future be reduced by the decisions of Smithfield Foods and Maple Leaf Foods – the largest pork producers in the US and Canada, respectively – to phase out sow stalls.

Of course, getting rid of sow stalls is only the beginning. It doesn't mean that pigs will be able to go outside, to roam around a pasture, or to have straw rather than bare concrete to lie down on. Even when sow stalls have gone entirely, there will still be a long way to go. But the readiness of voters, legislatures, and big corporate animal producers to make changes shows that animal suffering can be reduced, on a very large scale, by democratic, nonviolent processes. Obviously, as long as most people continue to want to eat animal products, a key role in these decisions is the demonstrated viability of alternative ways of meeting that demand. That is what the Food Animal Initiative is trying to achieve. When I toured their facilities at Wytham a few years ago, I was impressed by the significantly better quality of lives for the animals kept there than in the more conventional commercial operations I have seen over the years. Yet the farm at Wytham is a viable commercial operation, paying its own way without the assistance of sponsorships or research susbidies.

Many people will ask how we can really know what good animal welfare is. Marian Dawkins' aptly defines it as a situation in which "animals are healthy and have what they want." That raises the further question "How can we know what animals want?" The defenders of corporate agribusiness often say that their critics are responding "emotionally" to the sight of, say, six hens crammed into a small wire cage. A proper scientific approach, they say, indicates that, since the birds are laying eggs, their welfare is satisfactory. Here Dawkins has been a pioneer, finding ingenious ways of enabling the animals themselves to tell us what they want, and thereby showing that the science of animal behavior supports the critics of factory farming, and not its defenders.

This book is dedicated to two other pioneers in the struggle to give farm animals at least a minimally decent life: Ruth Harrison and David Wood-Gush. I would like therefore to take this opportunity to say that Ruth Harrison's *Animal Machines* had a huge influence on me when, as a graduate student in philosophy, I first began to think about the ethics of how we ought to treat animals. In 1970 *Animal Machines* was the only book to tell how animals were treated in the – then still relatively novel – factory farms that were increasingly providing more of the chicken, pork, and eggs I had unwittingly been eating. Ruth Harrison's powerful and well-documented attack on factory farming persuaded me that there could be no ethical justification for the way we were treating animals, and that if I wanted to have any respect at all for myself as an ethical person, I could not continue to eat animal products from factory farms. That set me on the path that led to *Animal Liberation*.

In an important sense, this book is continuing the work started by Ruth Harrison and David Wood-Gush, and bridging the gaps between science, farming, and the ethically concerned consumer. May this work continue to thrive.

Peter Singer
Princeton University

1 The future of animal farming

Roland Bonney and Marian Stamp Dawkins

Does animal welfare have a place in sustainable farming? Or does the threat of climate change now mean that the interests of nonhumans must be sacrificed to meet the demands of a rising human population? Can we improve the way we keep animals and still feed the world? Do we have to choose between ethics and economics? Between humans and animals?

The current debates about the future of the planet have thrown up answers to such questions that could be very bad news for farm animals. The Food and Agriculture Organization (2006), for example, has argued that since livestock contribute so much to global warming, the only solution is more and more intensive agriculture – animals packed closely together to make the maximum use of space and kept inside to limit the damage they can do to the environment. The Food and Agriculture Organization (FAO) foresees more selective breeding and more genetic manipulation to produce animals that survive under such conditions. So just at the point where more and more people are becoming concerned with the ethics of how their food is produced (Tudge, 2004; Singer & Mason, 2006), animal welfare is in danger of being pushed off the agenda.

The aim of this book is to challenge the "them and us" thinking that sets the interests of humans and farm animals against each other and to show that to be really "sustainable," farming needs to include, not ignore, the welfare of farmed animals. Animal welfare is so closely linked to human health and to the quality of human life that true sustainability cannot be a choice between economics and ethics or between human welfare and animal welfare. Sustainability must mean having it all – viable farms, healthy safe food, protection for the environment, as well as better lives for our farm animals.

But in today's global economy, with its increasing concern about climate change, is this possible? Isn't animal welfare a luxury for a rich minority and quite irrelevant to the majority of people in the world who cannot afford it? Surely there is not enough space, or enough money, or enough anything to achieve high standards of animal welfare when we are not even managing to ensure basic standards of human welfare? Surely we are going to have to make some very difficult choices. Of course we are. The point we want to make in this book is that those choices should put animal welfare at the heart of farming, even for those who put human welfare first. You don't have to care much about animals at all to see that their health and welfare will affect the health and well-being of you and your family and the whole human species through the food you eat, the diseases that might affect you, and the impact that agriculture of all sorts has on the whole planet.

As Bernie Rollin explains more fully in the next chapter, this interdependence between humans and animals can be seen as a kind of contract – a "deal" that goes back over thousands of years of human history. Traditionally, the deal was that farm animals provided us with food, clothing and much else while we provided them with food, protection from the elements and from predators. Humans have most often cared for their animals not out of sentiment but because their animals were valuable to them. With the industrialization of agriculture, we have broken that contract. Many people are no longer in touch with how farm animals are raised and so the health and welfare of food animals no longer seems to affect their own survival so directly. But indirectly it still does. Disease in food animals has potentially catastrophic effects on human health and the ecological effects of poor farming practices threaten the very life of the planet. It is time to renew the ancient contract for the benefit of all of us, not because it would be a pleasant extra if we could afford it but because it is a necessity we cannot afford to be without.

The exact terms of the new contract have yet to be worked out in detail because there are no easy solutions to the problems that confront us. On our side, there will have to be many changes – in our mind sets, in our diets, in our business models, and in the ways we keep animals. Furthermore, the future itself is uncertain as far as the technology that might become available or the changes in climate that might occur are concerned. But the essential elements are already clear. As this book shows, there are successful ways of farming that give priority to animal welfare, deliver high quality food, protect the environment and, most importantly, make business sense. What people value in their food is changing and continues to change. As a result, there are some surprising changes in the way that global businesses set their priorities. There are commercial as well as social and ethical benefits to animal welfare.

The scale of the challenge

Agriculture is the largest industry on the planet. It employs 1.3 billion people and provides a livelihood for about a billion more (Food and Agriculture Organization, 2006a). Livestock production, which uses land both for grazing and for growing animal feed, takes up 30% of the ice-free land on the planet and is responsible for 18% of all greenhouse gas emissions measured as CO_2 equivalent – higher even than transport. Of the gases emitted as a result of human activity, livestock are responsible for 37% of the methane, 65% of the nitrous oxide, and 64% of the ammonia. Much of this pollution comes from manure but livestock have an even more insidious effect on the environment. Livestock now account for about 20% of the total animal biomass in the world and 30% of what they now occupy was once the habitat for wildlife. Through forest clearance, livestock farming could thus be said to be the biggest destroyer of biodiversity. In Latin America, for example, 70% of previously forested land is now occupied by pasture (Food and Agriculture Organization, 2006a).

And that is just the current situation. The human population of the world is currently 6.5 billion. It is growing by 76 million each year and 95% of this population increase is in developing countries. The UN's medium projection forecasts that world population will reach 9.1 billion by 2050. Not only is the populations growing , so are individual incomes and, as people become richer their demand for food and other agricultural products also increases. With higher disposable incomes, people move towards more varied diets that include more pre-processed food, more foods of animal origin and more added sugar and fat (Food and Agriculture Organization, 2006a). The statistics are staggering. Currently 276 million tonnes of meat is being produced globally each year, increasing by approximately 3% each year. The UN's Food and Agriculture Organization's prediction is that global production of meat is set to double from its value at the beginning of this century to over 465 million tonnes in 2050. That means that the environmental impact of the animals that are currently farmed must be halved just to keep environmental damage to its current level.

Increasing demand for animal food products exerts extra pressures on an economy in all sorts of ways. It encourages advanced breeding and feeding technology in livestock production and it leads to the development of irrigation systems and the use of fertilizers to increase production of plant crops on which to feed the animals. However, the largest pressure is for change in the very structure of livestock production itself. Much animal farming is rapidly shifting away from extensive systems towards more intensive, "landless" production, particularly of pigs and poultry. Comparisons between world livestock production systems by the FAO (averages for 1991–1993 and 2001–2003) show that globally, 14.6% of total meat is produced in grassland-based systems (ruminants), compared to 33.6% in landless systems. Total pig meat output rose by 30% at world level, an increase accounted for almost entirely by Asia. The total production of poultry meat grew by approximately 75%, again with the highest expansion in Asia. By contrast, cattle stocks are up 5% and small ruminants 10% (Steinfeld et al., 2006).

All of us, rich or poor, city dwellers or rural farmers, are affected by agriculture and by the rising global demand for animal products. Against this background of potential water shortages, pollution and environmental damage, rising fuel costs, and the need to feed the human population, it would be all too easy for animal welfare to get lost. Indeed, the FAO response to what it calls "Livestock's Long

THE FUTURE OF ANIMAL FARMING

Shadow" is to concentrate entirely on the human issues and to see the future of farming in terms of intensive, indoor units, which have the aim of maximizing productivity and limiting the environmental damage food animals can do. There is no mention of animal welfare anywhere in the report (2006). The aim of this book is to redress that balance and to explore a more optimistic future for animal farming.

The scope of the book

We will be questioning the pessimistic view that there are just two possible futures for animal farming: either more and more intensive farms or no meat eating at all. Many different people will argue from many different points of view that these are not the only alternatives in front of us. We want to show that there is another future that involves farming in a sustainable way but also makes sure that food animals have reasonable lives. The basis for this optimism is the fact that there already are successful commercial farms that are putting into practice many ideas that could form the future for animal farming if enough people want them and are prepared to make them work. The contributors to the book come from a diversity of backgrounds – from big business, from animal welfare organizations, from academic institutions, and from practical farming. They certainly do not agree with each other on everything but two common threads unite them. They all agree that farm animals *matter* and they all agree that sustainable farming must have animal welfare at its core, along with healthy food, human welfare, and environmental protection.

The book falls loosely into two sections, the first making the case for why major changes in animal farming are necessary, the second describing what the changed farming will be like. However, the fact that there is no clear distinction between the case for change and change that is already happening is one of the most encouraging features of the book. It shows that the aspirations are not only realistic but practical.

The book is dedicated to two people who directly or indirectly inspired many of the contributors to this book. Ruth Harrison's book *Animal Machines* (1964) awoke the public to the dangers of intensive agriculture and "factory farming." Throughout her life Ruth continued to campaign on behalf of farm animals and was never afraid to make herself unpopular on their behalf. David Wood-Gush was Professor of Agriculture at Edinburgh University and brought veterinary science and ethology (the study of behavior) into the service of animal welfare. His "family pig" system has now, with modification, been adopted with great success into a commercial venture and, as we will see throughout the book, provides a case study for how it is possible to bring about sustainable change in animal farming. We remember them and thank them.

Part 1
Voices for change

We start with the arguments in favor of changing the way in which animals are currently farmed. Some of these arguments come from looking at how farm animals are kept now and the deep-seated conviction that many people have that all is not well. This imperative for change comes from a rejection of current farming practices because of the harm they do to animals (Rollin, Chapter 2; D'Silva, Chapter 3; and Midgely, Chapter 4 , who argue that we must change because farm animals suffer from what is now done to them). But other arguments reach the same conclusion from quite different starting points. Even if current farming methods are considered acceptable as far as the animals are concerned, we must change because what we are doing is unsustainable. We cannot continue as we are because we will run out of space, food, and water and we will be overcome by disease and pollution. This imperative for change comes from the fact that current farming methods cannot last. The threat of climate change has seen to that.

It is the convergence of these two kinds of argument – concern for animals and the need for sustainability in the face of climate change – that is now bringing about the possibility of real change (Rawles, Chapter 5). Public concern for animal welfare may not in itself be enough to have a major impact on farming as a whole. But couple it with the need to farm in a way that helps not hinders the problems of what is happening to the climate and we have a powerful and unstoppable force for change in farming.

2 The ethics of agriculture: the end of true husbandry

Bernie Rollin *is one of the leading voices on the ethics of the way in which humans treat animals. He is an academic philosopher at Colorado State University but has a uniquely practical approach to ethics. His belief that philosophers should address serious applied issues and not just play intellectual games can be seen in his books such as* Animal Right and Human Morality *and* Farm Animal Welfare.

Animal agriculture

The subject of this chapter is a betrayal of great magnitude, the modern human abrogation of our ancient contract with animals and with the earth, which contract nourished and sustained the growth of our civilization and, with consummate irony, allowed us to develop the science and technology which in turn enabled us to cavalierly disregard that same contract. This betrayal is not only a moral violation of our age-old relationship with animals, but a prudential denial of our own self-interest. For unless we renew the symbiosis intrinsic to that contract, we will be unable to nourish ourselves physically; in the end, like all animals, we must eat to live, reproduce, and survive. And the end of this contract means an end to a renewable food supply, without which abrogation cannot be sustained.

This ancient contract, which we will characterize as husbandry with regard to animals, and stewardship with regard to the land, is the bedrock upon which economics, art, and culture rests. Yet with the profound hubris of an Icarus who challenged inherent human limitations, with blind and abiding faith in the humanly crafted tools which repeatedly show themselves as impotent in the face of natural disaster, we thumb our noses at both morality and prudence. As the ancients crafted the tower of Babel, so we overreach the constraints imposed on us by the natural world.

These seemingly lofty and pretentious claims will be fleshed out in the body of this chapter, in depressing detail. But it is fitting to tether our discussion to a pair of anecdotes experienced by the author that will point the way to that discussion.

The examples evoke both our balanced past and our tenuous future in the area of animal agriculture.

About three years ago, I was visiting a rancher friend in Wyoming, and having dinner at his home along with a dozen other ranch people. I asked the dinner guests how many of them had ever spent more money on medical treatment for their cattle than the animal was worth in economic terms. All replied in the affirmative. One women, a fifth generation rancher, asked, with something of an edge, "What's wrong with that, Buster?" I replied "Nothing from my perspective. But if I were an agricultural economist, I would tell you that one does not spend $25 to produce a widget that one sells for $20." She fairly spat her reply: "Well that's your mistake, Buster. We're not producing widgets, we're taking care of living beings for whom we are responsible!"

Virtually every rancher I have encountered – and I have lectured to around 15,000 across the US and Canadian West – would respond in a similar vein. Even if they do not spend cash, ranch people often sit up all night for days with a marginal calf, warming the animal by the stove in the kitchen, and implicitly valuing their sleep at pennies per hour! Children of ranch families often report that the only time their father ever blew up at them was when they went to a dance or a sporting event without taking care of the animals. These ranchers represent the last large group of agriculturalists in the US still practicing animal husbandry, as we shall explain shortly.

In contrast to this elevating anecdote, consider the story told to me by one of my colleagues in Animal Science at Colorado State University. This man told of his son-in-law who had grown up on a ranch, but could not return to it after college because it could not support him and all of his siblings. (Notably, the average net income of a Front Range (i.e.eastern slope of the Rocky Mountains) rancher in Colorado, Wyoming, or Montana is about $35,000!) He reluctantly took a job managing a feeder pig barn at a large swine factory farm.

One day he reported a disease that had struck his piglets to his boss. "I have bad news and good news," he reported. "The bad news is that the piglets are sick. The good news is that they can be treated economically." "No," said the boss. "We don't treat! We euthanize (by dashing the baby pigs' heads on the side of the concrete pen)." The young man could not accept this. He proceeded to buy the medicine with his own money and clock in on his day off, and treated the animals. They recovered, and he told the boss. The boss's response was "You're fired!" The young man pointed out that he had treated them with his own time and money, and was thus not subject to firing. He did, however, receive a reprimand in his file. Six months later he quit and became an electrician. He wrote to his father-in-law: "I know you are disappointed that I left agriculture, Dad. But this ain't agriculture!"

In these two stories is encapsulated the history of where agriculture was traditionally, and what it has overwhelmingly become today. How and why did this change take place?

The traditional account of the growth of human civilization out of a hunter-gatherer society invariably invokes the rise of agriculture, i.e. the domestication of animals and the cultivation of crops. This of course allowed for as predictable a food supply as humans could create in the vagaries of the natural world – floods, droughts, hurricanes, typhoons, extremes of heat and cold, fires, etc. Indeed, the use of animals enabled the development of successful crop agriculture, with the animals providing labor and locomotion, as well as food and fiber.

This eventuated in what Temple Grandin refers to as the "ancient contract" with animals, a highly symbiotic relationship that endured essentially unchanged for thousands of years. Humans selected among animals congenial to human management, and further shaped them in terms of temperament and production traits by breeding and artificial selection. These animals included cattle – dubbed by Calvin Schwabe the "mother of the human race" – sheeps, goats, horses, dogs, poultry and other birds, swine, ungulates, and other animals capable of domestication. The animals provided food and fiber – meat, milk, wool, leather – power to haul and plow; transportation; and served as weaponry – horses and elephants. As people grew more effective at breeding and managing the animals, productivity was increased.

As humans benefited, so simultaneously did the animals. They were provided with the necessities of life in a predictable way. And thus was born the concept of husbandry, the remarkable practice and articulation of the symbiotic contract.

"Husbandry" is derived from the Old Norse words "hus" and "bond"; the animals were bonded to one's household. The essence of husbandry was care. Humans put animals into the most ideal environment possible for the animals to survive and thrive, the environment for which they had evolved and been selected. In addition, humans provided them with sustenance, water, shelter, protection from predation, such medical attention as was available, help in birthing, food during famine, water during drought, safe surroundings, and comfortable appointments.

Eventually, what was born of necessity and common sense became articulated in terms of a moral obligation inextricably bound up with self-interest. In the Noah story, we learn that even as God preserves humans, humans preserve animals. The ethic of husbandry is in fact taught throughout the Bible; the animals must rest on the Sabbath even as we do, one is not to seethe a calf in its mother's milk (so we do not grow insensitive to animals needs and natures); we can violate the Sabbath to save an animal. Proverbs tells us that "the wise man cares for his animals." The Old Testament is replete with injunctions against inflicting unnecessary pain and suffering on animals, as exemplified in the strange story of Balaam

THE ETHICS OF AGRICULTURE: THE END OF TRUE HUSBANDRY

who beats his ass, and is reprimanded by the animal's speaking through the grace of God.

The true power of the husbandry ethic is best expressed in the 23rd Psalm. There, in searching for an apt metaphor for God's ideal relationship to humans, the Psalmist invokes the good shepherd:

> The Lord is My shepherd; I shall not want. He leadeth me to green pastures, He maketh me to lie down beside still waters, He restoreth my soul.

We want no more from God than what the good shepherd provides to his animals. Indeed, consider a lamb in ancient Judaea. Without a shepherd, the animal would not easily find forage or water, would not survive the multitude of predators the Bible tells us prowled the land – lions, jackals, hyenas, birds of prey, and wild dogs. Under the aegis of the shepherd, the lamb lives well and safely. In return, the animals provide their products and sometimes their lives, but while they live, they live well. And even slaughter, the taking of the animal's life, must be as painless as possible, performed with a sharp knife by a trained person to avoid unnecessary pain. Ritual slaughter was, in antiquity, a far kinder death than bludgeoning; most importantly, it was the most humane modality available at the time.

The metaphor of the good shepherd is emblazoned in the western mind. Jesus is depicted both as shepherd and lamb from the origin of Christianity until the present in paintings, literature, song, statuary, and poetry as well as in sermons. To this day, ministers are called shepherds of their congregation, and pastor derives from "pastoral." And when Plato discusses the ideal political ruler in the *Republic*, he deploys the shepherd–sheep metaphor: the ruler is to his people as the shepherd is to his flock. Qua shepherd, the shepherd exists to protect, preserve, and improve the sheep; any payment tendered to him is in his capacity as wage-earner. So too the ruler, again illustrating the power of the concept of husbandry on our psyches.

The singular beauty of husbandry is that it was at once an ethical and prudential doctrine. It was prudential in that failure to observe husbandry inexorably led to ruination of the person keeping animals. Not feeding, not watering, not protecting from predators, not respecting the animals' physical, biological, physiological needs and natures, what Aristotle called their *telos* – the "cowness of the cow" the "sheepness of the sheep" – meant your animals did not survive and thrive, and thus neither did you. Failure to know and respect the animal's needs and natures had the same effect. Indeed, even Aristotle, whose worldview was fully hierarchical with humans at the top, implicitly recognized the contractual nature of husbandry when he off-handedly affirmed that though the natural role of animals is to serve man, domestic animals are "preserved" through so doing. The ultimate sanction of failing at husbandry – erosion of self-interest – obviated the need for any detailed ethical exposition of moral rules for husbandry: anyone unmoved by self-interest is unlikely to be moved by moral or legal injunctions! And thus one finds little written about animal ethics and little codification of that ethic in law before the twentieth century, with the bulk of what is articulated aimed at identifying overt, deliberate, sadistic cruelty, hurting an animal for no purpose or for perverse pleasure, or not providing food or water.

Though this ancient contract with domestic animals was inherently sustainable, it was not in fact sustained with the coming of industrialization. Husbandry was born of necessity, and as soon as necessity vanished, the contract was broken. The industrial revolution portended the end of husbandry, for humans no longer needed to respect their animals to assure productivity. In symbolic advertisement of the breaking of our sustainable contract with animals, in the mid-twentieth century academic departments of animal husbandry in the US became departments of animal science, with industry replacing husbandry. Indeed, textbooks of animal science characterize the field as "the application of industrial methods to the production of animals." The values of productivity and efficiency replaced the values of husbandry, to the detriment of animals, sustainability, the environment, agriculture as a way of life, rural communities, stewardship, and a respectful, moral stance towards the living things we built our civilization on.

Animal husbandry may be characterized as putting square pegs in square holes, round pegs in round holes, and creating as little friction as possible in doing so. Failure at husbandry meant that one's animals did not produce; failure to respect animal needs and natures hurt oneself as well as the animals. This was suddenly overridden by technological means, as it were "sanding tools," that allowed us to force square pegs into round holes, and round pegs into square holes, and where, at least in the short run, productivity flourished at the expense of respect for animal needs and natures.

Consider, for example, the egg industry, one of the first areas of agriculture to experience industrialization. Traditionally, chickens ran free in barnyards, able to live off the land by foraging and express their natural behaviors of moving freely, nest-building, dust-bathing, escaping from more aggressive animals, defecating away from their nests, and, in general, fulfilling their natures as chickens. Industrialization of the egg industry, on the other hand, meant placing the chickens in small cages, in some systems with six birds in a tiny wire cage, so that one animal stands on top of the others and none can perform any of their inherent behaviors, unable even to stretch their wings. In the absence of space to establish a dominance hierarchy or pecking order, they cannibalize each other, and must be "debeaked," producing painful neuromas since the beak is innervated. The animal is now an inexpensive cog in a machine, part of a factory, and the cheapest part at that, and thus totally expendable.

The steady state, enduring, balance of humans, animals, and land is lost. Putting chickens in cages and cages in environmentally controlled building requires large

THE ETHICS OF AGRICULTURE: THE END OF TRUE HUSBANDRY

amounts of capital and energy and technological "fixes"; for example to run the exhaust fans to prevent lethal build-up of ammonia. The value of each chicken is negligible so one needs more chickens; chickens are cheap, cages are expensive so one crowds as many chickens into cages as is physically possible. The vast concentration of chickens requires huge amounts of antibiotics, vaccines, and other drugs to prevent wildfire spread of disease in overcrowded conditions. Breeding of animals is oriented solely towards productivity, and genetic diversity – a safety net allowing response to unforeseen changes – is lost. Small poultry producers are lost, unable to afford the capital requirements; agriculture as a way of life as well as a way of making a living is lost; small farmers, who Jefferson argued are the backbone of democracy, are superceded by large corporate aggregates. Giant corporate entities, vertically integrated, are favored. Manure becomes a problem for disposal, and a pollutant, instead of fertilizer for pastures. Local wisdom and know-how essential to husbandry is lost; what "intelligence" there is is hard-wired into "the system." Food safety suffers from the proliferation of drugs and chemicals, and widespread use of antimicrobials to control pathogens in effect serves to breed – select for – antibiotic-resistant pathogens as susceptible ones are killed off. Above all, the system is not balanced – constant inputs are needed to keep it running, and to manage the wastes it produces, and create the drugs and chemicals it consumes. And the animals live miserable lives, for productivity has been severed from well-being.

One encounters the same dismal situation for animals in all areas of industrialized animal agriculture. Consider, for example, the dairy industry, once viewed as the paradigm case of bucolic, sustainable animal agriculture, with animals grazing on pasture giving milk, and fertilizing the soil for continued pasture with their manure. Though the industry wishes consumers to believe that this situation still obtains – the California dairy industry ran advertisements proclaiming that California cheese comes from "happy cows," and showing the cows on pastures – the truth is radically different. The vast majority of California dairy cattle spend their lives on dirt and concrete, and in fact never see a blade of pasture grass, let alone consume it. So outrageous is this duplicity, that the association was sued for false advertising and a friend of mine, a dairy practitioner for 35+ years, railed against such an "outrageous lie."

In actual fact, the life of dairy cattle is not pleasant. In common with many other areas of contemporary agriculture, animals have been singlemindedly bred for productivity, in the case of dairy cattle, for milk production. Today's dairy cow produces three to four times more milk then 60 years ago. In 1957, the average dairy cow produced between 500 and 600 pounds of milk pass lactation. Fifty years later, it is close to 20,000 pounds (Colorado Dairy Industry, 2005; USDA/NASS, 2006). From 1995 to 2004 alone, milk production per cow increased 16%. The result is a milkbag on legs, and unstable legs at that. A high percentage of the US dairy herd is chronically lame (Nordlund et al., 2004) (some estimates range as high as 30%), and these cows suffer serious reproductive problems. Whereas, in traditional agriculture, a milk cow could remain productive for 10 and even 15 years, today's cow lasts slightly longer than two lactations, a result of metabolic burnout and the quest for ever-increasingly productive animals, hastened (in the US) by the use of bovine somato-trophin (BST) to further increase production. Such unnaturally productive animals naturally suffer from mastitis, and the industry's response to mastitis in portions of the US has created a new welfare problem by docking of cow tails without anesthesia in a futile effort to minimize teat contamination by manure. Still practiced, this procedure has been definitively demonstrated not to be relevant to mastitis control (see Bagley, 2003). (In my view, the stress and pain of tail amputation coupled with the concomitant inability to chase away flies, way well dispose to more mastitis.) Calves are removed from mothers shortly after birth, before receiving colostrum, creating significant distress in both mothers and infants. Bull calves may be shipped to slaughter or to a feed lot immediately after birth, generating stress and fear.

The intensive swine industry, which through a handful of companies is responsible for 85% of the pork produced in the US, is also responsible for significant suffering that did not affect husbandry-reared swine. Certainly the most egregious practice in the intensive swine industry and possibly, given the intelligence of pigs, in all of animal agriculture, is the housing of pregnant sows in gestation crates or stalls – essentially small cages. The *recommended* size for such stalls,* in which the sow spends her entire productive life of about four years, with a brief exception we will detail shortly, is 3 feet high by 2 feet wide by 7 feet long – this for an animal that may weigh 600 or more pounds. (In reality many stalls are smaller.) The sow cannot stand up, turn around, walk, or even scratch her rump. In the case of large sows, she cannot even lie flat, but must remain arched. The exception alluded to is the period of farrowing – approximately three weeks – when she is transferred to a "farrowing crate" to give birth and nurse her piglets. The space for her is no greater, but there is a "creep rail" surrounding her so the piglets can nurse without being crushed by her postural adjustments.

If kept outside with access to suitable materials, a sow will build a nest on a hillside so that excrement runs off. She will forage an area covering a mile a day; and take turns with other sows watching piglets and allowing all sows to forage (Rollin 1995). But in a confined stall, the animal's nature is frustrated, she effectively goes mad, exhibits bizarre and deviant behavior such as compulsively chewing on the bars of the cage. She also endures foot and leg problems, and lesions from lying on concrete in her own excrement.

These examples suffice to illustrate the absence of good welfare in confinement. Rest assured that a long litany of additional issues could be addressed. In general, all animals in confinement agriculture (with the exception of beef cattle who live most of their lives on pasture, and are "finished" on grain in dirt feed lots,

---

*   Editors' note: Such stalls are banned in the European Union.

THE ETHICS OF AGRICULTURE: THE END OF TRUE HUSBANDRY

where they can actualize much of their nature), suffer from the same generic set of affronts to their welfare absent in husbandry agriculture.

1   *Production diseases* – By definition, a production disease is a disease that would not exist or could be of serious epidemic import were it not for the method of production. Examples are liver and rumenal abscesses resulting from feeding cattle too much grain, rather than roughage. The losses from animals that get sick are more than balanced by the remaining animals' weight gain. Other examples are confinement-induced environmental mastitis in dairy cattle and "shipping fever." There are text books of production diseases, and one of my veterinarian colleagues calls such disease "the shame of veterinary medicine" because veterinary medicine should be working to eliminate such pathogenic conditions, rather than treating the symptoms.

2   *Loss of workers who are "animal smart"* – In large industrial operations such as swine factories, the workers are minimum wage, sometimes illegal, often migratory workers with little animal knowledge. Confinement agriculturalists will boast that "the intelligence is in the system" and thus the historically collective wisdom of husbandry is lost, as is the concept of the historical shepherd, now transmuted into rote, cheap, labor.

3   *Lack of individual attention* – Under husbandry systems, each animal is valuable. In intensive swine operations, as illustrated earlier in the anecdotes illustrating industry versus husbandry, the individuals are worth little. When this is coupled with the fact that workers are no longer caretakers, the result is obvious.

4   *The lack of attention to animal needs determined by their physiological and psychological natures* – As mentioned earlier, "technological sanders" allow us to keep animals under conditions that violate their natures, thus severing productivity from assured well-being.

Animal welfare issues do not exhaust the ethical concerns occasioned by industrial animal agriculture. There are numerous other major societal concerns:

1   *Environmental* – Obviously running industrial agricultural operations requires a great deal of fossil fuel both to run the operations and to move the products great distances. Furthermore, such operations generate enormous amounts of manure. Unlike the valuable role of manure in pastoral agriculture, where it nourishes the soil, in confinement, manure disposal becomes a huge problem with some giant swine operations I have visited generating as much excrement as a good-sized city, as one official (proudly) told me. Excess manure leaches into ground water and pours into surface water under conditions of high rain, as once occurred in North Carolina. The wastes in turn produce

significant odor, and eutrophication of streams, rivers, and lakes, i.e. growth of undesirable algae and bacteria. In the central valley of California between San Francisco and Los Angeles, many giant dairies have generated unprecedented air pollution consisting of organic volatile compounds, nitrous oxide, ammonia, and methane, eliciting unprecedented environmental regulations.

2   Closely connected to environmental contamination are *human health issues*. Two thirds of human infectious diseases are zoonotic, and close confinement allows infectious microorganisms to burn through populations, much like a cold in a dormitory. In addition, crowded conditions are conducive to high volumes of pathogens mutating rapidly. When antibiotics are used as a sanding tool to compensate for unhealthy conditions, or as a growth promoter at low levels, a crucible for creating antibiotic-resistant bacteria is generated. Worker health then becomes a problem, both because of pathogens and because of bad air. In some swine barns, workers must wear respirators, though the animals do not! And the air pollution mentioned earlier in the Central Valley of California is responsible for marked increased incidence of respiratory disease, cardiovascular problems, and pre-natal and neonatal health problems.

3   *Loss of small agriculture and destruction of rural communities*. As mentioned, in some 26 years the US has lost 87.8% of the swine producers operating in 1980 (Vansickle, 2002), with the hogs now produced by large companies. From over a million producers in the 1960s, by 2005, the number had fallen to 67,000 (USDA/NASS, 2005). As the small hog farmers have gone out of business, the once thriving communities they nurtured have become ghost towns. This in turn kills the communities. And in rural areas where large operators have become established, major cultural conflicts occur between traditional inhabitants and the migratory workers. In the face of these considerations, we must again recall Jefferson's admonition that small farms and farmers are the backbone of democracy – no one wishes to see major corporations monopolizing the food supply.

4   *Food safety* – Transporting animals over great distances, giant slaughterhouses, mixing of animals from different areas, and farming for productivity and efficiency have led to major food safety issues. *E. coli* H5707 is one such example. The most dramatic example is the dreaded BSE – bovine spongiform encephalopathy – which most likely came from feeding meat and bone meal to cattle; natural herbivores.

5   "Externalized costs" – What helped drive industrialized agriculture's evolution is the desire for "cheap food." Americans spend only 9% of their income on food, as opposed to the 20% spent by Europeans. But it should be clear from our discussion that what one pays in the supermarket does not represent the true cost of animal products created by industrial methods. There are human

THE ETHICS OF AGRICULTURE: THE END OF TRUE HUSBANDRY

health costs (in addition to the suffering associated with illness), for example pollution from dairies in the Central Valley of California costs every man, women, and child in that area an estimated 3 billion dollars, or $1000 per year in direct medical costs (Hall et al., 2006). The costs of environmental pollution and the clean-up it will eventually require are inestimable. And how does one cost-account the animals' suffering?

As consumers become aware of these issues, increasing numbers are "voting with their dollars" to buy from nonintensive producers who provide good animal welfare and practice environmentally sound, sustainable management. The growth of specialty markets and restaurants in the US who carry these products and buy locally is far more precipitous than that of traditional supermarkets and restaurants. And in Britain and the European union, consumers demand has forced chain markets such as Tesco to impose environmental and welfare requirements on their suppliers. In addition, in Northern Europe and in the EU, legislative constraints have been placed on industrial agriculture.

The history of the concept of animal welfare provides a fitting conclusion to our discussion of industrialized animal agriculture. Under husbandry, the notion of animal welfare was philosophically transparent. An animal was well-off if the physical and psychological needs flowing from its nature – what I elsewhere call *telos* – were met. Under such conditions, good productivity was an excellent criterion for welfare – only animals that were thriving would grow properly or give milk or eggs. Thus the issue of conceptual analysis of "welfare" did not command much attention. As industry replaced husbandry, the old concept was taken for granted to obtain – if the animals were productive, said animal scientists, they were of necessity well-off. (This is manifest in the 1981 definition of welfare enumerated in the Council for Agricultural Science and Technology (CAST) report on farm animal welfare generated by the industrial agricultural industry, where welfare was so defined (CAST, 1981).)

This, of course, is a conceptual error. Equating productivity with welfare only works in conditions of true husbandry. Here productivity and welfare are extensionally equivalent (i.e. cover the same set), if not the same in meaning. But, in industrial agriculture, meeting the animal's needs flowing from their *telos* is not a necessary condition for productivity – the "technological sanders" we discussed override the need for respecting animal nature. Thus productivity today is in no way an assurance of well-being.

But, as European legislation has shown, and US ethical evolution indicates, citizens in western society wish to preserve the fair contract with animals implicit in husbandry, and eroded by industrialization. "If we consume animals, they should at least live decent lives" is easily elicited from most people, and buttressed by surveys. So a new concept of animal welfare was required. In the early 1980s,

Marian Dawkins (Dawkins, 1980), Ian Duncan (Duncan, 1981), and this author (Rollin, 1981) contributed to the concept by pointing out that welfare was in large measure a matter of the animals' subjective experiences – the way in which the conditions in which they are kept matter to them. This was a great conceptual leap beyond productivity only reluctantly accepted (if accepted at all!) by producers and scientists, but certainly self-evident to ordinary common sense (i.e. that animals have thoughts and feelings).

The last component needed for a full analysis of animal welfare is the ethical dimension. Given that husbandry no longer obtains, we must not only worry about the animals' experiences entailed by meeting or not meeting their needs, but must also ask which needs and to what extent we ought to meet them. Industrial producers worry only about the physical needs required for productivity, and only to the extent that such fulfillment meets maximal efficiency for production. Hence the feeding of bone meal, sawdust, chicken manure, etc. to farm animals provides nourishment at minimal cost but with no respect for animals' natures (such as whether they are herbivores), or about their physical health except to the extent that it impacts on production. In today's society, the new definition of welfare that is implicitly emerging as a societal ethic is that in raising animals for food, we are obliged to worry about all needs emerging from their *telos* we can practically meet (e.g. freedom of movement, social needs). Thus, whereas a producer would accept a productive animal as well-off if it were producing milk, for example, even if it were painfully lame, ordinary citizens would not accept an animal, however productive, living in constant pain! As Europe has demonstrated, the emerging societal ethic decrees that if this no longer occurs automatically as presuppositional to production, it should be mandated by regulation or legislation.

Crop agriculture

Just as technology replaced animal husbandry with an industrial approach, so there has also been a loss of land stewardship. In barely half a century, the agriculture of symbiosis and sustainability has been superseded by a technocratic mindset that says, in essence, we can do better than working with nature – we can shape it to better suit our needs.

As children, many of us learned about balanced aquariums. If we wished to keep a fish tank where the fish lived and we didn't want to keep tinkering with it, we needed to assure that the system in question was as close to a "perpetual motion" machine as possible, a system that required little maintenance because all parts worked together. That meant including plants that produced oxygen and consumed carbon dioxide, enough light to nourish the plants, plants that thrived in the available light source, water that was properly constituted chemically, scavengers

THE ETHICS OF AGRICULTURE: THE END OF TRUE HUSBANDRY

to remove wastes, and so on. When such a system worked, it required minimal maintenance. If something were out of balance, plants and animals would die, and require constant replacement. The fish tank aims at being a balanced ecosystem, and thus represents a model of traditional approaches to cultivation of land, wherein one sought to grow plants that could be grown indefinitely with available resources, which conserved and maximized these resources, which would not die out or require constant enrichment. Hence the beauty of pastoral agriculture, where pasture nourished herbivores, and herbivores provided us with milk, meat, and leather, and their manure enriched the pasture land in a renewable cycle.

Cultivation of land evolved locally with humans. If one did not attend to the constraints imposed by nature on what and how much could be grown in a given region, the region would soon cease to yield its bounty, by virtue of salinization, or depletion of nutrients or overgrazing, or insect infestation. Thus, over time, humans evolved to, as one book put it, "farm with nature," which became like animal husbandry, both a rational necessity and an ethical imperative. Local knowledge, accumulated over long periods of trial and error, told us how much irrigation was too much; what would not grow in given soils; what weeds left standing protected against insects; where shade and windbreaks were needed, and so on. Thus accumulated wisdom was passed on – and augmented – from generation to generation, and was sustainable, i.e. required minimal tweaking or addition of resources. The genius of agriculture was to utilize what was there in a way that would endure. As in animal husbandry, if the land did not thrive, you did not thrive. Traditional agriculture, then, was inherently sustainable; by trial and error over long periods of time it evolved into as close to a "balanced aquarium" as possible.

As with animal husbandry, humans broke their contract of stewardship over the earth as soon as technology gave them the tools to do so. No longer was agriculture directed at sustainability. Instead it was driven by a mantra of productivity: suck as much yield from the earth as possible with the help of technology, sometimes in the name of profit and sometimes in the name of "feeding the world." This meant huge tractors for tillage, irrigation systems, fertilizers, pesticides, herbicides. What had historically been gentle intercourse with the earth became rape and plunder. And yield grew.

But at what cost? Science and technology enabled extraction of greater crop yields than hitherto imagined, but, as is said, there is no free lunch. Instead of farming depending on idiopathic wisdom, local knowledge, know-how – passed from generation to generation – it was now conceived of as technology, as applied science, as nomothetic – law like – and in principle applicable to any locale. As in animal agriculture, capital and machinery supplanted knowledgeable labor; farms got bigger and bigger, "get big or get out" became their mantra. Food was plentiful, and thus cheap.

With these gains came major costs, albeit costs that were not immediately obvious and long term, but costs they were. If you forget about having a balanced aquarium, then you must pump resources in regularly to compensate for the loss of balance. The new agriculture required a great deal of fuel, to run the machinery and to make the chemicals. Massive amounts of water were also required. Only 75% of modern agriculture's consumption of water is replenished. And these chemicals, fertilizers, pesticides, herbicides, and fuels left residues that polluted air and water and lead to disturbance and death of fragile ecosystems. Growing crops required land, so forests and rain forests were cleared, uprooting and impoverishing endogenous ecosystems, annihilating species, losing wild plants whose significance to the balance of nature was unappreciated. Repetitive tillage led to soil erosion, and depletion, degradation, and pollution. Toxic residues abounded. Excess irrigation depleted soil nutrients, requiring yet more fertilizer. Monoculture – cultivation of one crop, that is the most profitable, and most productive, but also most susceptible to devastation (by putting all of one's eggs in one bucket) – replaced traditional crop diversity.

Powerful and plentiful agricultural chemicals had a negative health effect on workers and citizens. Indiscriminate use of pesticides predictably and inexorably led to inadvertent selective breeding of super-pests, highly resistant to these chemicals – even as the massive use of antibiotics in confinement animal agriculture both to promote growth and to mask the effects of bad husbandry led to the evolution of antibiotic-resistant pathogens. In addition, pesticides nonselectively killed off both desired pests, and their natural enemies.

As farmer debt load increased, increasing numbers of small farmers lost their farms, unable to afford the infrastructure required. And as they went out of business, so the little communities they inhabited died and, with them, their culture and their way of life. And, with unsurpassed and bitter irony, in the US, the land grant universities chartered in the nineteenth century to help small farmers and the rural communities they constituted contributed to their demise, by developing the very technologies that caused the problems we have described, funded to do so by the United States Department of Agriculture (USDA). Farming as a way of life became agribusiness, the grand monoliths dominating the food supply that Jefferson feared as inimical to democracy. And agribusiness funded the science that perpetuated agribusiness, leaving no niche for the small farmer. So bad has this become that when I did contract research for the USDA on farm animal welfare and cited western rancher views of their animals, and the sort of concern for their well-being we evidenced earlier, I was told by a high official that "western ranchers are not real agriculturalists – they are a bunch of damn Romantics!"

In the end the restoration of husbandry to animal agriculture and stewardship to crop agriculture both embody prudential and ethical imperatives. The prudential dimension is simple – if we fail to create sustainable systems, we will eventually be unable to raise animals or cultivate the land, in the absence of affordable (in all senses) inputs. The ethical dimension incorporates our obligations to the future of humanity and to renewing our ancient contract with the animals.

3 Why farm animals matter

Mary Midgley *is a philosopher who has always worried about the strained relations that arise between humans and the other species around them. Her books include* Beast and Man *and* Evolution As A Religion. *Her chapter here raises the question why people find it so hard to notice how the animals they eat are treated, so that ethical queries about farm animals have lagged far behind those about other animals.*

Some animals are much more equal than others

Why do farm animals matter? Well, no doubt for the same reason that other animals matter – namely, that things can matter to *them*. Pigs and cows mind what happens to them in a way in which cars or even cabbages do not, and this minding seems to have moral consequences. For one thing, it puts us within range of the Golden Rule – not to do to others what one wouldn't like to have done to oneself. Most people, when actually confronted with a lamb or a flotilla of ducklings, are likely to see this aspect of things at once and they will often conclude that the way in which we treat these creatures matters. But at other times, when they are not so confronted, they are liable not to think like this at all. Pâté de foie gras may then seem perfectly in order. Thus the interesting question here is perhaps – How is it that we behave as if they did not matter? Or indeed, Why have we taken so long to see that they do?

Human life contains many such incongruities, many paradoxes of this kind. They all have their histories, and some elements of this one are quite recent. For instance, at the Festival of Britain in 1951 there was an exhibit celebrating the British Hen. The walls were lined with egg-cardboard, to rejoice in its victory over paper bags, and the whole thing was a paean of triumph about the splendid new machinery – standard cages and so forth – which had, at last, turned the chicken industry into a fully efficient production-line. This was part of the mechanistic euphoria which followed the end of the war – the celebration of new technologies which were expected to give us a new and entirely error-free future. In that context it was not at all surprising that farm animals quietly stopped being fellow-creatures and mutated into "agricultural products" – which was, of course, until lately, their official status in the European Union.

Of course that status was not wholly new. Probably throughout our history, for as long as animals had been domesticated at all, humans have looked at them with a kind of squint – one eye seeing them simply as things or "products," the other as something more or less like people. Practical treatment of them, too, has oscillated wildly between these two poles. Often they were killed or abused without hesitation, simply as a matter of course. At other times, selected creatures were treated with great consideration, quite as much consideration indeed as was given to some humans. A really interesting case is the parable which the prophet Nathan tells to King David. As Nathan explains:

> ... The poor man had nothing save one little ewe lamb, which he had bought and nourished up, and it grew up together with him and with his children; it did eat of his meat and drank of his own cup, and lay in his bosom and was unto him as a daughter ... And ... the rich man took the poor man's lamb and [killed it and] dressed it [to feed a guest]
>
> And David's anger was greatly kindled against the man, and he said to Nathan, "As the Lord liveth, the man that has done this thing shall surely die." (Samuel, Chapter 10)

The story has, of course, a hidden point. Nathan tells it as a fable to entrap David by showing its likeness to a recent mean and brutal action of his own, and he will thunder in reply, "Thou Art the Man." But the striking thing is the way in which the prophet can assume that the story itself is shocking. Neither David nor Nathan doubts that the poor man might reasonably love his lamb in this way – that he could quite properly buy it in order to add it to his family, or that the rich man has committed an atrocious crime. And this even though the whole thing takes place in a context where it was assumed that all Hebrew sheep destined to end up on someone's table. Just so, in the parable of the Good Shepherd, the shepherd lays down his life for his sheep. But the actual literal sheep continued to end their lives in the traditional way.

What are the features that can lift animals out of thinghood in this way and bring them within the charmed circle of human consideration? In various cultures, simply rearing a young animal can have this effect. That process tends to turn them into companions (Serpell, 1986), no doubt because these situations arouse in us the kind of attention that we would normally give to young children, which naturally leads to increased awareness of their feelings. Some domestic animals which work with people or share much of their lives, notably horses and dogs, are also seen in this way as companions, because they too are regarded with extra interest and attention. Often, however, they are attended to very selectively, getting consideration in some contexts and not in others.

Sometimes, like cats in Egypt, certain creatures are cherished because they are regarded as sacred. Rather differently, too, hunted animals, especially large ones, have often been glorified and respected. This may happen because in their case too – though very differently – hunters have to attend to them carefully enough to become aware of their feelings. That does not, of course, save them from being killed, but it often means that they are honoured, somewhat unfairly, for their independent, wild status. They then tend to be contrasted with farm animals – ox, pig, fowl – which are despised as inferior beings, the boring, ignoble bottom rung of the human hierarchy, rather than being welcomed as exotic guests from outside it.

Cartesian automata

Thus, in our culture, human–animal relations have blundered on for most of history in a kind of unexamined ambivalence where widely different attitudes arose, or could be deliberately called up, very much according to convenience or the mood of the time. Some philosophers – Plutarch, Montaigne, Voltaire, Bentham, Schopenhauer, Mill – grew angry about the resultant injustice and tried to make people attend to it (Singer & Regan, 1976). But the dominant European tradition treated it as a minor issue which could safely be left in confusion. When these muddled habits of mind needed to be expressed and defended, the usual explanation given was that animals did not matter because they had neither reason nor souls.

Great stress was, of course, laid on Reason – that is, on the connection between thoughts rather than on the thoughts themselves. European philosophers, from the Greeks on, were so much occupied in trying to make people think more clearly that they tended to exalt reason itself as the crucial human attribute, and to center reasoning on the human attribute of speech. Animals, who use other ways of communicating, therefore came to seem increasingly alien. And, about souls, the Christian tradition had no doctrine of rebirth which could have related animals more closely to human life. They were officially regarded as dispensable, "beasts that perish," beings opposite in every way to human souls destined for eternal glory.

It might have been expected that the Enlightenment would have shifted these attitudes to animals by making people think more clearly about them. And in the end that did happen. It was the Enlightenment that produced that great surge of general concern about suffering called the Humanitarian Movement, which launched campaigns against slavery and also, eventually, campaigns on behalf of animals. But this concern about suffering was a slow and painful development which only became effective in the late eighteenth century. Long before that, in the 1640s, Descartes had laid down the deadly doctrine that animals were genuinely unconscious automata – machines, like the clockwork figures which so impressed thinkers of that age that they saw them as the prototype for the whole cosmic system. Human bodies too were deemed to be machines, but ones blessedly animated by conscious spirits inside them. Humans were thus the only conscious beings in the universe.

WHY FARM ANIMALS MATTER

As the Industrial Revolution rolled on and machinery became an increasingly familiar part of people's lives, this dualistic way of thinking appeared more and more natural. The image of mind and body as distinct beings chimed neatly with the industrial one of workers operating machines. Mechanistic theory had, of course, a laudable aim. It was meant to liberate science – in particular, physics – from entanglement with human motivation, allowing it to study the material system in appropriate terms. Inevitably, however, it encouraged callousness towards animals, and this may not have been merely an accident. Physiologists were beginning to see a great future for research by dissection, and were usually prevented from doing it on people. Harvey's discovery of the circulation of the blood, which deeply impressed Descartes, was a triumph of mechanistic thinking and seemed likely to herald many more such successes. Experimenters therefore eagerly dissected live animals – of course without anesthetics, since none had yet been invented. They explained that, as Descartes' doctrine made clear, this work was not really cruel because their subjects were quite unconscious. Any noises such as screams and groans that might emerge were merely grinds and creakings of their mechanism. (Today, these are called "vocalizations".)

This was the point when the idea began to spread that science itself demanded readiness to ignore the feelings of animals. In the nineteenth century the great physiologist Claude Bernard expressed this view strongly. Scientists, he said, should never listen to complaints that lay people might raise about their methods because these were matters internal to science. No outsider could understand them. Scientific objectivity (said Bernard) meant something more than fair and balanced argument. It meant that the items one was examining were themselves simply *objects* – mere inert matter with no point of view of its own. And many scientists took up this strange conviction. As humanitarian sentiment grew stronger in the general public this intransigence naturally lent great bitterness to controversy about experiments on animals – a bitterness which still persists today.

By contrast, the situation of farm animals attracted much less attention. Here there was no sudden change comparable to the advance of physiology, no new, sectional interest in exploiting them. But then, nothing new was needed. Farm animals went on being exploited (of course on a far wider scale than experimental ones) for the same ancient, basic and widely shared reason that they always had been; they were wanted for food. Certainly the idea that they need not be eaten had sometimes been floated. Even in the ancient world there were vegetarians. But these were mainly minority groups, ascetic believers in religious systems, and the reasons they gave for abstention mostly concerned ritual purity and especially avoiding the defilement of blood rather than animal suffering. The Pythagoreans backed these points by ideas of reincarnation, and Plutarch did express a direct horror of cruelty itself. So did Montaigne. But it was not until the rise of humanitarianism in the late eighteenth century that that message began to be widely heard.

At that time, people seem to have become aware, by some great new imaginative effort, of some of the suffering that they had been inflicting. And, remarkably, these reformers extended that awareness – as earlier political reformers had not – to include the lives of other animals as well as people. Thus Jeremy Bentham, besides campaigning against the savagery of the penal system, also attacked the inhuman treatment of animals. Wilberforce, besides campaigning against slavery, also denounced brutality towards the beasts. And Tom Paine cried out that animals as well as humans were being oppressed. These were the ideas that inspired vegetarians such as Blake and Shelley, and they led to a variety of campaigns conducted throughout the nineteenth century, which did shift public opinion enough to produce various laws making cruel practices such as bull-baiting illegal. That progress, however, remained slow and patchy because the debate became polarized in a way that often made sensible controversy difficult. People campaigning on behalf of animals tended to be dismissed as unbalanced eccentrics.

Enter the psychologists

One reason for this, obviously, was the general difficulty that affects all large reforms. Exploiting animals was so useful that people found it hard to imagine how they could possibly stop doing it. Besides this, however, there was a remarkable factor peculiar to the animal debate, namely the belief, just mentioned, that there was actually something specially scientific about treating animals as if they could not feel.

Looked at directly, this is certainly a strange proposition. Descartes' original thesis that animals were unconscious was not a piece of science but a philosophical speculation. He adopted it in order to rule that they differed from human beings in exactly the way required by his dualistic theory. He never supported it by any empirical evidence. It was just a device to bracket off these awkward borderline cases. And after his time the development of biology gradually disclosed how wild his speculation had been. Research showed that the human nervous system is very similar to those of other species, especially to those of the great apes – creatures whose very existence was unknown in Descartes' time – so similar that research on one was constantly found relevant to the other. Eventually, when Darwin made it clear that humans are actually descended from animals much like those that now surround us, the continuity between our lives and theirs became clearly visible. The dualist suggestion that they were entirely alien no longer made any sense. What is interesting is how many people still managed not to see this.

Darwin himself grasped the continuity clearly. In his book *The Expression of the Emotions in Man and Animals* (1872) he used examples from humans and other species together as a matter of course, carefully noting the differences that

WHY FARM ANIMALS MATTER

actually appeared, but in general finding – as today's ethologists have also found – striking and fascinating parallels. And in writing about these he used the familiar language of human social life, adjusting it carefully, but without any embarrassment, to make his acute and fertile observations.

Darwin simply ignored the unreal scepticism which had long insisted that, even if animals have feelings, we could never know anything about them. This was, of course, related to the wider view that we cannot know about other people's feelings either. He showed the way out of both these sceptical pits, into which so many theorists had unnecessarily dug themselves, by concentrating on expressive behavior and our innate ability to read it. As he pointed out, we, along with other social animals, absolutely need to know the feelings of those around us, and we often need this just as much with other species as with our own. Recent thought has followed him here, and – delightfully enough – the discovery of mirror neurones, which provide the mechanism for it, has made the idea acceptable to many people who would previously have thought it unscientific. They could not, it seems, accept what had been happening to them all their lives until they had been shown how it worked in a laboratory.

Darwin was, however, one of the last people who was allowed to think like this. Soon after his death, this branch of psychology abruptly entered an ice-age where any suggestion of sympathy with animals – any nuance insinuating that they could have a mental life comparable in any way with that of humans – became "anthropomorphism" and amounted to professional suicide. Direct interest in them was considered sentimental and degrading to a scientist. Edward Thorndyke, an influential psychologist of the day, beautifully displayed the distant tone that was now required when he wrote, "My first research was in animal psychology, not because I knew animals or cared much for them, but because I thought I could do better than had been done."

The horror of anthropomorphism; Lloyd Morgan's Canon

In the context of such pronouncements the word *anthropomorphism* is itself interesting. Originally it was a theological term, a name for the heresy of supposing God to have a material human body. In the nineteenth century, however, it was seized on to describe a quite different kind of sin – namely, speaking of other animals in the same terms that are used about human beings. This word at once imported a metaphysical flavour – a suggestion of something grotesquely unsuitable, even blasphemous – into very ordinary contexts, such as the proposal that a cow might have maternal feelings towards her calf or a sow towards her piglets. Thus the word served to crystallize and justify the researchers' impression that an immense, Cartesian difference between humans and other species should be taken for granted.

No doubt that is why it did not strike Thorndyke that ignorance about animals might not actually be an advantage for somebody who was researching on them. His attitude is not surprising since psychologists were not now interested (as Darwin had been) in the animals themselves but were using them purely as simplified models for human psychology. Not wanting that simplicity to be blurred by irrelevant details they carefully designed their methods to exclude these. They therefore ruled in Lloyd Morgan's Canon (1878) that researchers must "never interpret an action as the outcome of a higher psychical faculty if they could possibly interpret it as produced by a lower one" (cited in Burchfield, 1975). Animals, that is, must never be given credit for understanding what they did if their actions could possibly be explained in some other way – by habit, error, conditioning, or anything else that did not involve thinking. Thus the burden of proof was always placed on any view that departed from the default position – that animals are profoundly insensitive and stupid.

Morgan's canon has usually been seen as a principle of parsimony. Yet a little thought shows that it is not parsimonious at all. To assume, as a default position, that animals – all animals – are profoundly stupid is not a cautious, minimal stance, suitable for a prudent enquirer. It is an extreme one. Neither is it disinterested, since it obviously flatters human vanity. Nor is this in general an effective strategy for interpreting behavior. If, in human affairs, we never consented to believe that other people understand what they are doing until we had carefully excluded all other possible motives that might cause their actions, the business of life would not go on well.

This applies, too, to dealings with other animals as well as with humans. There are many people who depend on the willingness and intelligence of animals, such as horses and dogs for their work. These are not sentimentalists but practical individuals such as horsemen, shepherds, farmers, and circus-workers. They do not find it necessary to be constantly sceptical on the point. When necessary, they readily entrust their lives to the judgment of their animals without constantly suspecting them of being secretly stupider than they appear. Generally speaking, in fact, intelligent action is its own sufficient warrant. Doubts only arise about it when actual symptoms of stupidity appear.

Considering all this, there is surely something odd about the determined refusal of psychological researchers to accept intelligent behavior at face value – a refusal displayed in Lloyd Morgan's canon and still maintained to some extent today. Whence comes this unshakeable conviction that apparent intelligence in animals is usually a cover for stupidity? The obvious source for it is Descartes' claim that animals are literally mindless because they have no immortal soul. But, now that

WHY FARM ANIMALS MATTER

the soul no longer forms part of modern scientific apparatus – now that Descartes has given way to Darwin – it is not at all clear why this radical division between humans and other animals should continue to be taken for granted.

Determining the cleverness or sensibility of particular animals is, for us now, normal empirical work, not metaphysical business as it was for Descartes. Questions about it have to be settled by observation and experiment. And by now, much of this has already been done. Ethologists, whose professional commitment does not involve *a priori* assumptions on the matter, have shown that animals can, in fact, do a great number of things that were traditionally supposed impossible for them, such as making tools or using mirrors. Each time when this happens, however, the discovery is viewed with suspicion as probably anti-scientific. For instance, pat on cue in a fairly recent issue of the *New Scientist* (December 16, 2006) there comes a report that researchers have discovered – surprise surprise – that monkeys can know when they don't know something (Phillips, 2006). The animals were provided with a "don't know" button and they quickly learned to use it intelligently. This feat, however, is supposed to be impossible for them because it involves "meta-cognition" – thinking about one's own thought – and this is not supposed to be possible without speech. Objectors complain that "metacognition has never been demonstrated in rats or pigeons – the workhorses of the psychology lab – and it is hotly debated even in primates. It is still often assumed that abstract thought needs some form of language."

Hierarchical problems

This is a more elegant way of saying what John Locke (1690) said ignorantly and crudely, "Brutes abstract not," but it still isn't true. Of course Locke was right that they don't use abstract nouns, but then we know already that they don't talk so that isn't very interesting. The researchers whom *New Scientist* reports make the very reasonable suggestion that developing general, structural concepts of this kind is not a consequence of using language but is an absolutely necessary pre-condition for it. Curiosity, for instance, is known to be widespread among intelligent animals. But how could one be curious about something without knowing that one did not know it? Human babies begin to be curious about the world around them long before they have any notion of language. Are they performing an amazing metacognitive feat, a feat impossible both to other animal babies and to adult humans? Or are they simply taking the normal first steps into a landscape that is common to all?

What makes it hard to think clearly about such questions is surely the habit of systematically downgrading animals in order to provide a clear, indisputable mark of human superiority. In recent times, theorists have tended to identify this mark as being a particular kind of consciousness or self-consciousness. This has a rather confusing effect because, not long ago, behaviorist orthodoxy sweepingly excluded the notion of consciousness from psychology altogether, for humans as well as for other animals. When this proved impractical for the human case it became necessary, either to take animal consciousness seriously along with the human sort, or to erect an emergency barrier between them by reverting to metaphysical dualism. Remarkably, it was the second course that was chosen. As Frans de Waal (2006) points out:

> The behaviorists' opposition to anthropomorphism probably came about because no sane person would take seriously their claim that internal operations in *our* species are a figment of the imagination ... Eventually the Behaviorists relented, exempting the bipedal ape from their theory of everything.
>
> This is where the problem for other animals began. Once cognitive complexity was admitted in humans, the rest of the animal kingdom became the sole light-bearer for Behaviorism ... Attribution of human-like experiences to animals was declared a cardinal sin. From a unified science, Behaviorism had deteriorated into a dichotomous one with two separate languages; one for human behavior, another for animal behavior.
>
> So, the answer to the question "Isn't anthropomorphism dangerous?" is that, Yes, it is dangerous to those who wish to uphold a wall between humans and other animals ... In the end we must ask; *What kind of risk are we willing to take, the risk of underestimating animal mental life or the risk of overestimating it?*

That is surely a good question. It is really not easy to see now why one of these risks should have been regarded with such horror while the other was largely neglected.

The ethological contribution

More recently, information from a different source has deeply altered our view of the balance between these two kinds of risk. At the time when the Behaviorist ice-age was still suppressing psychological speculation about animals, a different scientific approach to them was developing in another part of the forest, a part so deeply wooded that civilized people had never really noticed it before. Ethologists – people who, unlike Thorndyke, had a direct interest in studying animals – had begun to develop methodical ways of observing their behavior in the wild. Konrad Lorenz, Niko Tinbergen, Jane Goodall, Diane Fossey, and others devotedly investigated that hidden world, devising methods that were unmistakeably scientific, yet could in principle be understood by a wide public. Very soon, a tribe of skilled photographers was recording their discoveries on film for the rest of us to see. Television brought them into our homes, showing us things that we, in our urban existence, had formerly never heard of.

This was how, for the first time, many of us began to get some idea of the richness and variety of animal lives. The psychologists' vision of the standard abstract animal – a white rat or pigeon, shut up in a small cage for life – dissolved as we saw something of the huge range of experience that exists among creatures who actually have their living to make out in the world. The variety of these lives amazed us, yet we did not find them altogether strange. We saw, too, many common elements that resonated with our own experience, varieties of family affection, play, ingenuity, puzzlement, anxiety, rivalry, inner conflict that were close to our own lives. It turned out that, just as Lorenz said, nonhuman animals may be less like us intellectually than we expect, but they are often more like us emotionally. Since we are, after all, earthly animals ourselves it is not surprising that we often mind about things in the same sort of ways that they do. And it is this minding that brings them within range of our concern.

Accordingly, in the last half-century, some sympathetic awareness of animal lives has spread to a much wider public than it ever did before. That sympathy no longer seems totally eccentric, and it is gradually having an effect on people's toleration of various practices. At first, indeed, this awareness was concentrated chiefly on distant and exotic animals such as chimpanzees or lions. It has been slower to reach the less exciting creatures, such as farm animals, who are directly affected by our normal way of life. For instance, television companies systematically refused for a long time to show programs about the details of intensive farming because they thought them too unpleasant, even though films of predators killing zebras and wildebeest were accepted, like other killings, as normal evening fare. Eventually, however, this veto crumbled and informative films on the subject have undoubtedly helped to fuel protests that were made against devices like calf crates and sow stalls, both of which have now been made illegal in the UK.

These films brought home to people something particularly grotesque about these and similar methods of intensive farming – namely that they don't just bring animals to death; they ensure that they have no real life at all before they go. Death does indeed happen to us all, and in an omnivorous society the actual death of food animals has long been accepted. But lifelong deprivation of absolutely everything that makes life worth living or even endurable is a rather different matter. And careful, scrupulous filming shows beyond question that that is exactly what happens with methods like these.

If systems of this kind were being introduced for the first time today, public opposition would probably not have allowed them to become established in the first place. Unluckily, however, their introduction passed unnoticed in the post-war spasm of technological euphoria that I described earlier, before the recent increase in interest and sympathy towards animals drew attention to the matter. There was then great enthusiasm for increasing the nation's food production and very little public interest in ethology. Economists and politicians called for intensive farming systems and compliant engineers were left to devise them, undisturbed either by public attention or by any kind of input from zoology.

The way forward

Remedying these anomalies may, however, be easier now than similar reforms have been in the past. It is clear today that we do not face a single drastic choice between "eating animals" and "not eating them." There is a huge range of choice available to us about how to treat them first, even if we do still eat them. The situation is like that over animal experimentation, where a similar realization is dawning that we do not have to choose between forbidding all experiments and accepting every method that is used at present. Today – even though, unluckily, a tiny minority of extremists continues to darken counsel on this subject – effective discussion about it now goes on between humane scientists and scientifically literate humanitarians who share the aim of ending bad practice, whether that practice is bad from the ethical or the scientific angle or indeed – as often happens – from both. Similarly over farm animals, it is now clear that farm-literate humanitarians can work together with humane farmers to get a much more tolerable quality of life for creatures who have long been most bizarrely neglected. We should all wish more power to their elbows!

WHY FARM ANIMALS MATTER

4

The urgency of change: a view from a campaigning organization

Joyce D'Silva *was Chief Executive of Compassion in World Farming from 1991 to 2005 and now works as Ambassador for CIWF. Her concern for the welfare of farm animals has its roots in her upbringing on a farm in Ireland, was strengthened by a period working in India, and by reading the autobiography of Mahatma Gandhi. She believes that if we all come to recognize that farm animals are capable of psychological as well as physiological suffering, our attitudes and behavior towards them will change radically.*

You're a bird flying high over the earth or may be you're just looking out the window of your last long-haul flight. But it's the bird's eye view you want. Carry on for long enough and you realize both the vastness of the earth and yet its limitations. There is so much ocean (the Los Angeles–Auckland flight is a good one for this), mountain ranges – maybe snowy, or just plain rocky – there's forest and there's desert. There's also farmland, which gets added to – temporarily – as forests are felled, and which decreases as the desert encroaches.

There are people down there, although, from where you are, you can't see the individuals, but you can see their footprint, their often foggy urban sprawls, maybe their efforts to fire a forest or dam a river. Soon there will be many more people, all needing homes and food. No future farming here nor there can operate in a geographical vacuum. Neither Britain nor the US nor Brazil nor Kenya nor China can farm in a totally unilateral way. And yet the first priority for those countries – and all countries – must be to feed their own people, as far as possible, from their own land.

Have I forgotten the farmed animals? No way. But their future is inextricably linked to questions about soil quality, water availability, environmental pollution, and global warming. Their future is linked to global policy on food, farming, trade, and the environment. In practical terms, it's imperative that farming nurtures the physical resources of our world – the soil, water, and air which are the wellspring of growth. It must nourish the soil, use the water wisely, and reduce as far as possible the noxious emissions which it produces. Yet it must operate under the imperative of feeding the earth's growing population. Ask any elderly Chinese person what it's like when over thirty million of your fellow humans die in famine around you.

Farming must meet humanity's basic needs for food and clean water. But sophisticated consumers don't just want their needs met – they often want more to eat than they need, they want a huge choice of products, and they want their food to be as cheap as possible, so that they have more money left for life's other necessities, or for pleasure pastimes. If they're in a rapidly developing country, they'll want to eat more meat and drink more milk, because that's equated with aspirational, wealthier lifestyles.

So where do the farm animals fit in to this complicated equation? If the cheap food paradigm is to be farming's only driver, then the animals haven't got a chance – no chance of living in any way which remotely promotes their well-being, their welfare. They'll be selectively bred to grow ever faster and meatier, or to produce more milk or eggs per year. They may be genetically modified and/or cloned to meet production targets even more quickly than hitherto possible. They'll be kept in the most confined space compatible with bare existence and they'll be denied basic comfort (– straw bedding *costs*). They'll be fed concentrated feeds which maximize growth and feed conversion ratios. If they misbehave, or if there's a chance they might misbehave – peck or bite each other – they'll have tails or beaks cut back. If there's a better slaughter price to be got on the other side of the world, they'll be sent there – alive. I call this kind of farming Fast-track Farming, although, traditionally, welfare groups have referred to it as "Factory Farming." Perhaps it epitomizes our fast-track age, when cross-global travel time has been cut to hours and global verbal and visible communication time has been cut to the blinking of an eye.

Is there an alternative way? I believe so. Farm animals could be bred back to more traditional, hardier breeds. They could be kept in farms which provide both comfortable shelter and the chance to range freely, as weather permits. They could be given feed appropriate to their species and left to seek some of their own food outside as their ancestors did. They could be kept in such good environments that they don't feel frustrated and competitive with their peers, so that they will be allowed to keep their bodies whole and intact. When the time comes for slaughter they could be taken quietly to a nearby slaughterhouse. I'm tempted to call this kind of farming Holistic Farming, but in case that's too alternative a description for you, let's call it Fair Farming. It encompasses the true husbandry that Bernie Rollin described in Chapter 2.

Let's look at these two types of farming and let's measure where we are now against them. Let's look for signals about the current direction we seem to be taking. Let's make an honest assessment of what kind of farming we want for the future – and how we might possibly achieve it. Let's assess what organizations and campaigns for better farm animal welfare have achieved so far and what challenges they face. Let's be clear too that campaigning organizations like Compassion in World Farming are fundamentally committed to the Fair Farming paradigm. I believe that the reasons for their commitment are ethical, sustainable, and scientifically sound.

Back in 1998, Compassion in World Farming organized a conference, "An Agriculture for the New Millennium – Animal Welfare, Poverty and Globalization." Although the majority of the speakers promoted Fair Farming concepts, Dennis Avery, Director of the Center for Global Food Issues at the influential American think-tank, the Hudson Institute, had some thought-provoking views to share, declaring: "The world must create five billion vegans in the next several decades, or triple its total farm output without using more land." Having dismissed the first option as much too difficult, Mr Avery went on to extol the virtues of intensification and genetic modification as the only way forward for global agriculture. Sadly, his vision for the future of animal farming epitomized the Fast-track Farming view. His best hope for the animals was that they "must be raised in carefully managed confinement facilities," where their productivity would be *tripled*.

What could this mean in real terms? How does Fast-track Farming already impact on the animals? Perhaps its prime operant is breeding. By selecting the fastest-growing, most productive animals over time, the world's breeding companies can provide animals with accelerated metabolisms and physiological changes: dairy cows with larger udders producing more milk, faster-growing chickens with proportionately more breast meat, cattle with meatier rumps, pigs with leaner muscle (meat), and hens who lay more eggs.

To welfare campaigners, this kind of selective breeding is the hidden scandal at the heart of Fast-track Farming because when you bring about such radical changes there are all sorts of side-effects which can be seriously detrimental to the animals' quality of life. The genetic selection of broiler (meat) chickens, for rapid growth and large breast development, has resulted in huge numbers of these birds developing pathological conditions of bones, joints, tendons, and skin. They are prone to painful hips and septic joints, and often develop sores, "hock burns," from lying down much of the time on the litter flooring of the shed which tends to become increasingly ammonia-ridden as the weeks go by.

This seems to me to be a tale of consequential suffering: bred to grow too fast, their legs become unable to support their body weight; selected for large breast growth they are tilted slightly forward, putting extra strain on the joints, which can become infected and arthritic; being in some pain, they will lie down for longer, thus giving extra opportunities for hock burns to develop from the ammonia leaching from the excreta-soaked litter on the floor. Similar problems affect the heavy strains of turkey in common use. John Webster (Emeritus Professor at Bristol University and former Head of the Veterinary School) concludes that

THE URGENCY OF CHANGE: A VIEW FROM A CAMPAIGNING ORGANIZATION

"approximately one quarter of heavy strains of broiler chicken and turkeys are in chronic pain for approximately one third of their lives" (Webster, 2004). In fact turkeys now grow to such heavy weights that natural breeding is impossible – the males could not mount the females without causing severe damage to both. All modern turkey breeding is performed by artificial insemination.

It's breeding like this which has facilitated the global chicken phenomenon: worldwide, we now rear and slaughter over 45 billion chickens every year, over 800 million of them in the UK. It's this which has brought the price of a kilo of chicken portions down in cost to the equivalent of a can of beer (www.tesco.com). It's this situation which inspires Webster (1994) to call it "in both magnitude and severity, the single most severe, systematic example of man's inhumanity to another sentient animal." Fast-track poultry farming has surely failed the animals.

There's another hidden horror behind today's fast-growing chicken. The broiler breeders – the birds who produce all those billions of chickens – have, logically, got an entirely different problem. They too have been bred for fast growth so that their offspring inherit this trait. But how then can they live long enough to reach puberty at 18 weeks or so and then go on living long enough to produce lots of eggs for hatching? How can their inbuilt fast-track growth rate be curbed? The answer is simple – and disgraceful. They are semi-starved. At various stages in their own growth, they are given reduced rations – sometimes only a quarter of what they would eat if fed ad lib. When Compassion in World Farming challenged the UK government on this in the High Court in 2003, the judge admitted that such birds were chronically hungry (whilst dismissing the case!).

Selective breeding has impacted cattle too. Dairy cows such as the ubiquitous black-and-white Holsteins have been bred to produce so much milk that they are metabolically being pushed to the limit. And there's no let up for the dairy cow. A couple of months after her first calf, she's pregnant again and continues to be milked until a few weeks before giving birth again, thus spending around eight months a year both pregnant and lactating. Frantic for food to sustain this level of output, the cow requires concentrated rations to supplement her diet, yet these same rations can be a factor in the appallingly high levels of lameness seen in cows, as the concentrated feed is acidic and can affect the feet, with 25% of cows requiring treatment for lameness every year. But not all cows get treated when lame. A survey showed foot damage in almost every cow at slaughter (Webster, 1994).

What could a GM future bring to cows? Already many high quality cows are induced, via hormones, to produce multiple oocytes, which are removed at the embryo stage and the embryos transferred to poorer quality surrogate mother cows. This does not usually require surgery, but the cow is subjected to human manipulation of her organs via her cervix. Although the UK requires an epidural anesthetic to be used when embryos are being inserted, not all countries do, in spite of the high probability of pain being caused.

Already the cloning of cows has become a reality, with ever more manipulation of source animals and surrogates likely, with the added welfare impact of cesarean deliveries of the cloned calves, who invariably grow to excessive sizes in utero. Some of these clones will also have been subjected to genetic modification, probably to alter milk composition or to increase milk yield. Yet again we play about with the actual genome of the animal, so that its body becomes our tool, designed to maximize every opportunity which may be physiologically possible, in order to augment our profits. Fast-track farming not only makes designer animals who grow too big and fast for their own good, it crowds and confines farm animals too. Space and land cost money. Even animals confined from a very young age, unable to turn round or walk, such as calves kept in narrow veal crates, can live long enough in these conditions to reach the desired slaughter-weight. They may stumble or even fall when they are taken out of their crates to be loaded onto the abattoir trucks, but they are still alive.

Hens kept five to a cage, each having less floor space than a sheet of typing paper, can produce eggs for around a year and could carry on for longer. But by the time these spent hens reach the slaughter house one survey showed 36% had broken bones (Gregory et al., 1990). Producing enough calcium to provide the shells for the 300+ eggs she lays each year, coupled with lack of exercise in the cage, has rendered the hen's bones fragile and brittle. In addition, the cage prevents the hen carrying out the regular sequence of things she would do in a natural environment – peck at the ground for food, stretch and flap her wings, dust-bathe to clean her feathers, fly up and perch at night (away from predators), and lay her eggs in a secluded nest.

Her cousins, the broiler chickens, aren't usually caged, but they too are horrendously crowded, often 17–19 of them to one square metre of floor space. As lameness becomes more prevalent and space is at an ever greater premium as the birds grow bigger, then, for some, it becomes difficult, sometimes impossible, to get to the food and water.

Intensive pig housing is also a prime example of Fast-track Farming, with the young pigs being fattened up in crowded concrete pens in a totally barren environment. Pigs are "switched on" to their highly sensitive snouts, which, in natural conditions they would use for exploring their environment and for rooting in the soil for roots and grubs to eat. With those two choices removed, rooting can transform into biting all too easily and young pigs, who love to indulge in play and dashing about, can be seen fighting and biting each other instead.

As Rollin has already described, mother pigs are sometimes kept for breeding in narrow metal-barred, concrete-floored pens – known as sow stalls – in which they are unable to turn round. Fast-trackers would claim this avoided aggressive behavior, reduced feeding costs (as immobilized sows have lower calorie requirements), and staff costs. As farrowing (parturition) time approaches, the sow is moved to a confined space, the farrowing crate, where she is virtually immobilized, and just given enough space to give birth and suckle her piglets, though not to nuzzle them. Her inherent instinct to build a safe and comfortable nest and defend it is thus ignored. You can see these sows making their nest-building movements with their snouts – down, forward, and up – but there are no leaves and twigs to build with and the whole exercise becomes a phantom performance.

I've not said much about sheep, as they have, with some exceptions, been spared the intensive indoor life. But they may suffer terribly from neglect and foul weather on remote hillsides, their lambs often die from hypothermia and they may be routinely tail-docked and the males castrated. For some strange reason, the law usually allows such mutilations to be performed without anesthetic on very young animals whilst often requiring it for older ones. I have yet to see proof that capacity for stress, pain, and fear grows with age. Sheep kept in vast extensive systems, such as the Australian outback, can also suffer from lack of care. The value of one sheep in many thousands is very small. For most farmers, the care or rescue of one animal is simply not cost-effective. Should a sheep trap a leg or develop a disease, they are unlikely to be spotted and may endure a painful and lingering death.

As for cattle reared for beef, their potential for good welfare may be high. The young calves often stay with their mothers and for much of their lives they may graze in fields. But two factors can undermine this idyll. Some, such as the Belgian Blue breed, are bred for double-muscling, which massively increases the amount of meat you can get from each animal. The downside is that cows giving birth to these brawny calves usually require cesareans. The other negative trend is to fattening up beef cattle either indoors, often on concrete slats which are slithery and uncomfortable, or in huge barren open-air feedlots, where they get a highly concentrated food ration and put on weight fast. Feedlots are common in the US and increasingly in Australia. Although outside, the animals are not able to graze, as there is only mud, not grass. Indeed these lots are stocked so heavily that any existing grass would inevitably get trampled into mud within a day. The cattle are fed a highly concentrated ration – and many may also have hormone implants inserted in their ears – all designed to speed up muscle growth, for muscle is meat. Shelter is not usually provided, so the animals may suffer the vagaries of the local climate, which may often be intensely hot. So there's fast-track beef too, much of it, no doubt, destined for the fast food industry.

So, although welfare organizations call for farm animals to be kept free range, i.e. outdoors, they do not mean outdoors at any cost. Animals should ideally be kept outdoors in an environment close to, or replicating, their natural environment. For cattle and sheep this would indeed be simply well-maintained pasture, which is not overstocked. For pigs and chickens this would mean light woodland or at least pasture with plenty of trees and bushes, providing shade for the pigs and shelter and perching opportunities for the chickens. In all cases, the animals should be subject to regular inspection, and if even just one animal is in pain, it should be treated, or, if necessary, humanely destroyed.

Feed itself is a huge welfare and environmental issue. Feed scientists continually look to develop the perfect ration to sell to farmers to encourage fast, healthy and productive growth. Some feeds will do just that. But common sense is sometimes forgotten in the search for the cheapest source of protein. If you had stopped someone in the street 20 years ago and asked them if they thought that feeding the remains of sheep and cattle to other cattle was a good idea, anyone knowing that cattle are natural herbivores would almost certainly have said "No." But it was cheap, it was done, and the consequences for cattle health and welfare have been disastrous, with cattle who developed BSE being destined for days if not weeks of disorientation and stumbling before being put out of their misery. As for the humans developing the associated new variant CJD, their agony, and that of their loved ones, can only be guessed at.

Other animals are fed particular feeds to induce speciality meats. Traditional white veal depended on feeding the calves a low-iron liquid diet throughout their lives, ignoring the calves' need for fibrous food in order for their ruminant multiple-stomach system to develop properly. By slaughter time, many of the calves would be bordering on a state of clinical anemia.

Geese and ducks reared for foie gras (fatty liver) suffer the horrors of having excessive quantities of fatty maize shoved down their throats every day in the weeks leading up to slaughter. Only by forcing in these unnatural quantities of food, will the birds' livers swell to the size required for this most cruelly produced product. It's a strange reflection on human nature, that these products produced by a most obviously cruel method are so highly valued by a small sector of society, who, perhaps, should know better.

So there's Fast-track Farming for you. A series of methodologies and practices designed to exploit the potential of different species of farm animals – at any cost. Never mind that the animals may never get to walk or fly, never get to live in natural social groups, that the young may never get nurtured by their mothers, or have only a highly abbreviated nurturing period, that daily life may be intensely boring – dare I say soul-destroying – and may often be a life engulfed in pain. Fast-track farming, factory farming, or whatever euphemism you wish to use, is based on the premise of utility: animals are there to be exploited to their maximum physiological potential in order that maximum profit can be had. There is no true husbandry.

Fast-track Farming is in denial of animals' capacities as sentient beings. It often fails to recognize their capacity for physical pain and suffering. It totally fails to acknowledge that farm animals, like their wild ancestors, have psychological

THE URGENCY OF CHANGE: A VIEW FROM A CAMPAIGNING ORGANIZATION

and social needs too. Calves want to be with their mothers – cows want to be with their calves; the same goes for lambs and ewes, piglets and sows. Young animals want to play; older animals often want to be in a family or social grouping of some sort. Fast-track Farming prides itself on being modern and up-to-the-minute in its use of the latest technology or feed ingredient or breeding method. In truth it is old-fashioned. It's old-fashioned because it hasn't kept up with new research showing the amazing range of farm animals' abilities. It has not embraced new research showing the capacity of chickens – and fish – to feel pain. It is in denial of research showing states of neuroticism in crated sows or bereavement-type behavior in cows deprived of their calves.

But let's be honest here too. The scientific world is also slow to attribute emotional states to farm animals. The literature is littered with "may indicate" and "is likely to be." The paradigm of the scientific method and the imperative of scientific credibility and reputation (and livelihood?) have, I believe, made many of those in the field reluctant to commit to too much. It is, after all, only 40 years since Jane Goodall was accused of anthropomorphism for naming the chimpanzees with whom she had been living and working for a year. They should have had numbers instead, she was told. Old attitudes die hard.

The truth can perhaps best be found in our own common sense – and in our hearts. If behaving towards farm animals with understanding and compassion is too hard to swallow, then I think we can at least agree to give them the benefit of the doubt in how we treat them. Let us assume that they feel pain and can suffer both physically and psychologically. Let us make this our premise when we design new systems and breeding methodologies. Let the well-being of the animals be our guide.

Some farming methods do take this field seriously. All over the world there are farmers – even companies – which put welfare high on their agendas. In the UK, organic bodies like the Soil Association keep animals in more sustainable groupings, minimize mutilations and transport times, and allow animals to range freely yet have decent, well-bedded shelter when needed. Welfare groups like the RSPCA, the Humane Society of the US, and many others give serious consideration to good welfare when they draw up rearing standards for farm animals, whose products they label accordingly.

Welfare groups and campaigning organizations have other important roles to play. It is they who have taken us inside the factory farms and, through their films, exposed the horrific truth to the public via the media. In 1995, CIWF's film of British calves being flown out from Coventry and trucked to a French veal crate farm where they were chained by the neck for the rest of their lives, not only caused outrage, but was used repeatedly on BBC television news whenever the topic of live animal exports was being introduced. TV programs have been made about farming practices, relying heavily on film footage from CIWF and other welfare groups. In such ways, the public learn about Fast-track Farming and can make decisions to avoid its products as far as possible. Welfare groups also produce referenced reports, make films for schools and produce a range of materials to be used in classrooms and lecture theatres. They translate these for use in other countries. Increasingly they develop good, often multi-lingual websites so that the whole world can become more informed and more motivated to take personal action. At the heart of their activities lies the drive to change the law, to win bans on cruel farming systems and practices, and to create minimum legally binding standards for farm animals. Political lobbying has burgeoned over the last 30 years, backed by public campaigning: marches, peaceful demonstrations, media events, petitions, and web-based campaigns. A string of successes has been achieved:

In 1984, CIWF's vigorous founder Peter Roberts, took a ground-breaking court case against a veal crate farm which was owned by a Priory in Sussex. The calves were unable to turn round and chained by the neck for life. The Deputy Prior told the media that he didn't know what all the fuss was about: "Animals can't suffer." UK law was inadequate at that time to sustain the prosecution and the case was lost. But the public outcry was massive. Within a year, the government announced it would ban the system and in 1990 it became illegal to crate calves unable to turn round and to feed them only on liquids. As with several other unilateral reforms, European Union (EU) reform has followed, often painfully slowly. In 2007 narrow veal crates were banned in the EU, although the standards laid down are poorer than the UK standards in several respects, allowing calves to be kept without bedding material such as straw, which is so important for calf comfort and dietary supplementation.

In 1990 CIWF persuaded Sir Richard Body MP to table a Private Members' Bill in the UK to phase out the keeping of pregnant sows in narrow sow stalls. CIWF and the RSPCA urged their members to write to their MPs and the Bill received such overwhelming support in Parliament that the government of the day felt compelled to take it on. An 8-year phase out was agreed and the system became illegal in the UK in 1999. As with veal crates, the EU followed suit, although the system will not be phased out until 2013 and some loopholes are left.

Both the RSPCA and CIWF have established European coalitions to enable pan-EU lobbying and campaigning to be effective. An interesting example of this dual – or perhaps multiple – pronged approach to EU campaigning occurred when the Directive on the welfare of calves was being discussed in 1996. The French government declared its absolute opposition to this Directive as it would mean the phasing out of the narrow veal crate, which was widely used in France. CIWF's new French office circulated thousands of postcards calling on the Agriculture Minister to change his mind. To everyone's amazement he did so, and the Directive was adopted. Only later did CIWF hear, from sources close to the Minister, that the reason for his change of mind was, "Well, he received so many postcards!"

THE URGENCY OF CHANGE: A VIEW FROM A CAMPAIGNING ORGANIZATION

One fundamental campaign was the brain-child of Peter Roberts. Disgusted with the wording in the Treaty of Rome, which categorized animals as "goods" or "products," in 1988 he began a campaign to have them recognized as "sentient beings," in other words, as creatures capable of suffering. CIWF was criticized, even within the welfare arena, for trying to achieve the unachievable – and who knew what "sentient" meant anyway? But with the assistance of groups in Europe, a million signature petition was handed in to the European Parliament in 1991 and endorsed by the Parliament three years later. CIWF continued to lobby government and Opposition, and a few weeks before the Treaty of Amsterdam was negotiated in 1997, the Labour party won the UK General Election. New Agriculture Minister Elliot Morley pressed for animal sentience to be on the agenda and was successful. A legally binding Protocol was added to the Treaty, recognizing animals as "sentient beings."*

Amazingly, within years of the start of this campaign, the term "sentient beings" was becoming common parlance in political speeches, academic research and even in the documentation of farmers' bodies like the National Farmers Union (National Farmers Union, 1995). Perhaps the greatest success of the pan-European strategy was in 1999 when the EU voted to phase out barren battery cages for laying hens by 2012. This could not have happened if support for the ban had come from just a handful of member states. But the southern European governments also came under pressure from welfare groups and success was achieved.

The European Commission itself has changed, allotting far more resources to farm animal welfare and taking the initiative in promoting action like the EU Community Action Plan on the Protection and Welfare of Animals. Meanwhile groups in the US were, on the whole, slow to take up the cause for farm animals, preferring to campaign on "safer" issues such as companion animals and wild animals. Around the late 1990s CIWF started engaging with leaders of the US groups and speaking at their conferences, urging them to campaign on farm animal welfare.

The new century has seen a dramatic change, with many US groups, both national and local, now majoring on factory farming issues. There have been successful ballot initiatives in both Florida to get rid of sow stalls (gestation crates as they are known in the US) and Arizona to ban both sow stalls and veal crates. New Zealand plans a phase-out of sow stalls similar to the EU, from 2015.

Now, however, there are new worries facing the European welfare groups. The EU now consists of 27, rather than 15 member states. Many of the accession countries have yet to develop an appreciation of the ways in which welfare is changing.

* "Protocol on protection and welfare of animals," in the Treaty of Amsterdam amending the Treaty on European Union, the Treaties establishing the European Communities and related acts. *Official Journal of the European Union* C 340, 10 November 1997.

So what strategies can the welfare groups adopt to deal with this new problem? Have they the resources to build up good partner groups in the offending countries? Their own resources are stretched already. But surely these countries cannot be ignored. CIWF believes that, whilst political lobbying must be maintained and widened geographically, now is the time to work with the food industry itself to raise its standards of farm animal welfare, where such exist, and to bring them into play where they are absent.

Since 2001, CIWF has run a bi-annual survey of the farm animal welfare standards of the major UK supermarkets, awarding "Compassionate Supermarket of the Year" to the winner. This has had the desired effect of getting the major chains to vie with each other for the top spot and for those lower down the chain to compete to avoid coming last. Since the survey started, Marks and Spencer (M&S) have gone free range in all shell eggs and egg product and Waitrose no longer sell battery eggs and are nearly nonbattery in (own-label) egg product. Other major chains are beginning to follow suit. Early in 2007, M&S announced further moves, going free range with fresh goose, turkey, and chicken and both they and Waitrose have reduced the stocking density of their intensive broilers.

Now, after engagement with CIWF's Food Policy Unit, not only are fast food chains like McDonald's going free range with all their eggs, but major companies such as Google and AOL have also decided to change their internal procurement policies to free range or at least nonbattery eggs. In 2007 CIWF awarded "Good Egg" status to several such companies. CIWF is extending its food policy work into the EU.

It would seem that the carrot and stick approach used by CIWF in its lobbying of both governments and the food industry is a strategy that works. Good practice is welcomed and congratulated, but all the players know that continued poor practice will be highlighted until reform is achieved. The burgeoning of Corporate Social Responsibility policies provides an ideal ethical space into which animal welfare can – and should – be incorporated. Already protecting the welfare of the animals reared for their products has become a publicized part of the CSR policies of companies like M&S and the new (American-based) player on the British scene, Whole Foods Market.

In the US, where change has lagged well behind the EU, burger chains have moved to enlarge the battery cages of their laying hens and, early in 2007, an amazing victory occurred when agri-business giant Smithfield announced it would stop its suppliers using sow stalls within 10 years. It seems that, after all, the food industry may begin to move faster than government. As welfare groups educate citizens, they become ethical shoppers, and supermarkets, restaurant chains, and food manufacturers have to begin listening. As welfare groups like CIWF and the RSPCA have increased their engagement with the food industry, it has felt compelled to respond. Already some of the big names in retailing and fast food are investing some of their profits in farm animal welfare research.

THE URGENCY OF CHANGE: A VIEW FROM A CAMPAIGNING ORGANIZATION

But, whilst the EU has moved dramatically on welfare reform, and industry is beginning to follow, other countries are fast-tracking at a remarkable speed. Indian hens are more likely to be caged than not and more and more Chinese pigs are being reared in brand new fast-track intensive farms. More and more broiler chickens are being bred for fast lives, probably destined for fast food. Meat production and consumption rates are also soaring. Forty years ago, the average number of kilograms of meat consumed per person per year was 56 kg in Europe – now it's 89 kg. In the US it was 89 kg, now it's 124 kg. But look to China and Brazil for the fastest growth rates: in Brazil up from 28 kg to 79 kg and in China up from 4 kg to 54 kg. In China, consumption rates in the cities are far higher and approaching European levels. Many of the rural population are still too poor to be more than occasional meat-eaters.

Already NGOs like CIWF, the RSPCA, and WSPA (World Society for the Protection of Animals) have started working in China, holding conferences, producing educational materials and developing good relations with policy makers. CIWF cannot support industrial animal agriculture systems, wherever they occur. The burdens they place on the health and welfare of farm animals are too great. We believe that no production system which uses fast-growing, high-yield breeds, encloses the animals in the minimum space, frequently mutilates their bodies and often feeds them unnatural types and amounts of feed can be sustainable in an ethical and holistic sense. Farm animals are not simply production units. Each creature is an individual sentient being with a capacity for well-being and, equally, a capacity for emotional and physiological suffering.

We have to find a better way forward for farming, one which incorporates the welfare paradigm at its core. As Colin Tudge persuades us in Chapter 16, we need an Enlightened Agriculture. This kind of farming will reduce the use of bought-in feeds which are such heavy users of artificial fertilizers and irrigation water, it will minimize fossil fuel energy on the farm, and it will stock only the numbers of animals whose waste the land can absorb and keep them in conditions which have high welfare potential. This kind of farming cannot rely just on the ethical stance of middle class, often western, consumers – it must be encompassed in national and global policy decision-making. It is the task of groups like CIWF to act urgently, so that when decisions are taken about the future of livestock farming – and taken they must be – animal welfare is built-in at every stage.

CIWF is striving for a world where meat production is reduced and consumption is based on products which come from animals who have lived lives of quality. We seek an agriculture where the environment is protected, where animal welfare is promoted, and where farming resumes its relationship with the farmer and the land. Such an agriculture is surely the only compassionate way forward for the future of farming (Food and Agriculture Organization, 2006a, 2006b).

5 Environmental ethics and animal welfare: re-forging a necessary alliance

Kate Rawles *was a lecturer in environmental philosophy for nine years at Lancaster University, before escaping to work freelance in 2000. She now works half-time as a lecturer in outdoor studies at the University of Cumbria and half-time as a freelance writer, consultant, and outdoor philosopher. She spent much of her childhood on farms, has lived and worked on a welfare friendly eco-farm, and has a longstanding interest in bringing together animal welfare and environmental concerns. Kate is a member of the Food Ethics Council, a Fellow of the Royal Geographical Society and of the Royal Society of Arts.*

Introduction

Modern, industrialized societies have brought extraordinary benefits to their citizens. They have also brought extraordinary costs to others – to other people, to animals, to other living things, and to ecological systems. In some respects they have brought costs to their own citizens, too. In relation to the environment, these costs are now so high they are jeopardizing our own futures, as well as those of millions of other species we share the earth with. Climate change, habitat degradation, and biodiversity loss are key issues here, though there are many others, and they are all interrelated.

How we think about the negative impacts of industrialized ways of life, and the relationships between people, animals, and the environment that underpin them, are explored in ethics and environmental ethics. For all their relatively low profile, these ethical issues are profoundly challenging. They reveal that the successes of modern lifestyles are built on attitudes towards others that are deeply unethical; and they raise really difficult questions about where we should go from here. They raise hard questions about the worldviews that underpin our actions as well as the actions themselves. In all of this, modern farming is an absolutely central issue. It is both a microcosm of the bigger picture and a major source of environmental and animal welfare problems in its own right. Questions about the future of farming are thus a central part of the wider challenge of figuring out how industrialized societies can move forward in an ethical and sustainable way.

There is no such thing as ethics-free farming. Farming by its very nature has relationships with and impacts on animals, on other living things, on ecosystems, on people, and on health. Whether explicitly or not, farming cannot help but take a position on what these relationships and impacts should be – on how these various "others" are to be treated. This means, of course, that there is no such thing as ethics-free food, either. However far removed many of us have become from agriculture, we all eat the products of farming, even if processed beyond recognition, two or three times a day. The future of farming therefore concerns everyone, or at least, everyone who eats, and we are all party to the ethics embedded in farming. In relation to mainstream, industrialized agriculture, this is not necessarily a comfortable place to be.

In this chapter, I will argue that a root cause of our environmental and animal welfare problems is an outdated and inappropriate worldview and ethic; that this inadequate way of seeing the world underpins many forms of industrialized agriculture; and that, in relation to animals and the environment alike, a significant shift in our worldview and ethics is urgently needed.

Animal welfare

Various other authors in this book have given powerful testimony to the negative impact of industrialized farming methods on animal welfare. These are not criticisms leveled at individual farmers, but at the inbuilt logic of the system that so many farmers are now part of. Like any major industry, the primary goal of modern agricultural systems is to maximize profit. One way in which this can be done is through economies of scale – by making farms larger and keeping more animals on them. Further economic gains can be achieved if these animals can be managed by fewer people. And further gains again are made if the animals are confined, so that less of the food fed to them is "wasted" by the animal moving around, and more is turned into meat or eggs. The end results are highly mechanized, industrial-scale systems that keep enormous numbers of animals in confined situations, managed by very few stock-people. Attendant animal welfare issues include severe reduction in behavioral repertoires, boredom, stress, social deprivation or social crowding, high levels of surgical and drug-based interventions, stereotypical behaviors, and other "vices" such as tail biting, as well as pain and fear (Singer, 1991).

Animals in these systems are viewed and treated as components in a production line. They are part of a process that aims to turn animal feed into human food as efficiently as possible. As Rollin and others have pointed out, various "advances" in modern farming methods have meant that we no longer have to understand and respect animals as sentient living beings to achieve this. The underlying ethic of this kind of farming endorses this treatment of animals as commodities or things rather than as living, feeling, experiencing beings. The "ancient contract" (Chapter 2) in which animals were, and had to be, treated at least reasonably well so that we could benefit from them, has been broken.

Farming and environmental issues

The notion of an ancient contract formerly based in necessity and now over-ridden, has relevance here too. Modern fertilizers and pesticides, for example, have meant that older ways of looking after the land in order to ensure continued soil fertility and pest control can be by-passed. And by-passed they have been. Of course, modern farming techniques have lead to greatly increased yields, at least in the short to medium term. But the bigger picture shows that this has been achieved at a very high cost indeed.

Ecosystem degradation and biodiversity loss

The Millennium Ecosystem Assessment (MEA) report makes for seriously depress-ing reading. Called for in 2000 by Kofi Annan, then the United Nations Secretary-General, and published in 2005, it represents the views of 1350 scientists from 95 countries. The report assesses the "health" of ecosystems and their ability to deliver "ecosystem services" to people. Its basic, stark message is that the ways in which human societies get their resources – food, water, timber, fiber, and fuel – are degrading the natural processes that support life on earth to a degree that raises serious questions about our own future (Millennium Ecosystem Assessment, 2005a).

> Human activity is putting such a strain on the natural functions of Earth that the ability of the planet's ecosystems to sustain future generations can no longer be taken for granted. (Millennium Ecosystem Assessment, 2005b)

> This report is essentially an audit of nature's economy, and the audit shows we've driven most of the accounts into the red (Lash, 2005)

Key factors in the degradation of ecosystems, according to the MEA Report, include habitat change, climate change, invasive species, over-exploitation of resources, and pollution such as nitrogen and phosphorus. In all of this, agriculture plays a critical role.

ENVIRONMENTAL ETHICS AND ANIMAL WELFARE: RE-FORGING A NECESSARY ALLIANCE

More land has been claimed for agriculture in the last 60 years than in the eighteenth and nineteenth centuries combined. An estimated 24% of the earth's land surface is now cultivated. Water withdrawals from lakes and rivers have doubled in the last 40 years. Humans now use between 40% and 50% of all available freshwater running off the land (MEA, 2005a).

In addition to the sheer scale of farming, there are impacts related to the particular ways in which modern farming is carried out. Nitrogen has become a major source of pollution, and more than half the synthetic nitrogen fertilizers ever used on the planet (they were first made in 1913) were deployed after 1985. Pesticides have had major biodiversity implications (MEA, 2005a). And the move towards large-scale, monocultural farms has lead to further reduction in available habitat and the possibility of farming coexisting with a diversity of other living things. To take just one example, 200,000 miles of hedgerow, a major farm habitat, have been taken down in the UK in the last 60 years. This is enough hedgerow to go around the world eight times.

The net result has been, according to the MEA Report, "substantial and largely irreversible loss in the diversity of life on earth, with some 10–30% of the mammal, bird and amphibian species currently threatened with extinction." (MEA, 2005a). Species, of course, have always gone extinct, but the current rate is estimated to be 100–1000 times faster than the natural, background rate of extinction. The loss of biodiversity has been more rapid in the last 50 years than any time in human history.

It is worth noticing that this alarming message comes not from environmental activists (though they have been transmitting this message for years) but from large numbers of respected scientists, working internationally, within the establishment, with no vested interests in the conclusions of their research. This of course makes the message all the more alarming. The same is true of the Intergovernmental Panel on Climate Change (IPCC) Report on climate change.

Climate change

The IPCC Report represents a degree of consensus across the international scientific community that is truly astonishing. Thousands of scientists agreeing!! It is against this background that the remaining few sceptical voices need to be understood. In the region of 2500 scientists involved in this report agree that climate change is happening, that it has a significant human cause and that it is very bad news. The details of exactly what climate change will mean, for different parts of the world, at different times are of course still under debate – but the big picture is clear (Intergovernmental Panel on Climate Change, 2007).

Global climate change is being driven by two main causes: the increase in the levels of "greenhouse gases" in the atmosphere, primarily through the burning of fossil fuels, and deforestation and other land use changes. Basically, humans have taken carbon that has been stored under the earth's surface in the form of coal, oil, and gas for millions of years and burned it, thus releasing CO_2 and other greenhouse gases into the atmosphere and exaggerating the otherwise natural and indeed life-enhancing greenhouse effect. At the same time we have taken down millions and millions of acres of forest ecosystems, and degraded other ecosystems that act as carbon sinks and that would help to counteract the increase in CO_2.

The end result is a problem, not just because of increasing temperatures, rising sea levels, and severe weather events, though these are of course important, but because of the degree to which weather patterns will change and, critically, because the change will be rapid. It is the speed of change as much as the change itself that people and other species will struggle to cope with.

The potential consequences of climate change for people and human societies are hard to exaggerate. They include the spread of diseases such as malaria and scarcity of food, water and land. The extreme weather and sea-level rise mean that millions will be affected by drought and flood (Intergovernmental Panel on Climate Change, 2007). According to the UK Meteorological Office, we can expect extreme drought to rise from the current 3% of earth's surface to 30% by 2100 (Burke et al., 2006). In other words, by the end of the century, almost a third of the earth will be uninhabitable with agriculture literally impossible. This will have its biggest impact in places like sub-Saharan Africa and is typical of one of the worst aspects of climate change – that although driven primarily by the richest people and countries, the worst impacts will be felt by the poorest. This in turn will lead to an increase in environmental refugees – 200 million, according to the Stern Report, which also predicts an impact on global economies equivalent to that of the last two world wars and the recession in the 1930s combined (Stern, 2006).

Of course, exactly how much warming and other changes we experience depends on what we do next. The IPCC presents a range of options from 1.8°C rise, widely accepted as the realistic minimum, to 6.4°C. To put these figures into perspective, the difference in average global temperature between now and the last ice age is around 4°C. Four degrees warmer than now is therefore very significant indeed. In fact, anything over 2°C has been classified as "dangerous." In relation to biodiversity, an increase of 1.5–2.5°C would put 20–30% of the species we share the earth with at risk of extinction (Thomas et al., 2004). An increase of 4.4°C would result in the actual extinction of more than half of all wild species of plants and animals (Intergovernmental Panel on Climate Change, 2007). As for the higher end of the IPCC spectrum, most analysts agree we really don't want to go there. As one summary put it, at +6.4°C "most of life is exterminated, humanity is reduced to a few survivors eking out a living in polar refugees." (McCarthy, 2007).

ENVIRONMENTAL ETHICS AND ANIMAL WELFARE: RE-FORGING A NECESSARY ALLIANCE

Climate change and farming

The relationship between farming and climate change is a complex one. Clearly, farming will be affected by changes in the earth's atmosphere. In some parts of the world, these effects will be positive, with longer anticipated growing seasons, for example, and the possibility of diversifying into new crops. Overall, however, the impacts will be negative with the speed of change again a critical factor, as well as the obvious problems for agricultural systems associated with droughts, floods, and storms. The limited number of species our farming systems depend upon makes agriculture especially vulnerable to change.

Farming is also a key contributor to climate change as well as being affected by it. Like any major industry, it is responsible for significant amounts of climate change related energy emissions associated with the construction and running of buildings and machinery, the transport of animals and produce, and so on. In addition, there are potent climate change gases associated with nitrogen-based fertilizers and with livestock. A recent report by the United Nations Food and Agriculture Organization, *Livestock's Long Shadow* (Food and Agriculture Organization, 2006a), has argued that livestock, primarily cattle, are responsible for nearly one fifth of the world's entire human-caused climate change emissions (Steinfeld et al., 2006). That's more than every plane, train, car, motorbike, and skidoo on earth (Rowlatt, 2007).

Root causes

Modern farming, then, is a major contributor to environmental problems. And these problems exist on such a scale that our own future is under threat. Climate change has captured attention in a way that ecological degradation and biodiversity loss have not – though some argue that the later are, in the end, the most serious environmental problems we face. But the message from both is clear. Industrialized societies, of which farming systems are a key part, are unsustainable. This way of living is promoted across the world as what it means to be developed, successful, progressed. Yet this way of living, and the values, ethics, and worldview that go with it, simply cannot be continued into the future without ecological collapse. It certainly cannot be shared by 6 billion people, let alone the projected 9 billion. As the WWF Living Planet Report puts it so powerfully, "if everyone enjoyed the lifestyle of the average Western European, we would need three planet earths" (World Wildlife Fund, 2004).

So where do we go from here? Steven Hawking has seriously suggested searching for other planets. Remaining earthbound, one approach is to try to technofix the problem – to increase efficiency and reduce waste within our production systems to such an extent that we *can* retain industrialized lifestyles, and share them, without causing ecological meltdown. Energy efficiency and waste reduction is certainly needed. Global energy consumption has gone up 80% in the last 30 years. Most climate change analysts argue that in industrialized countries, climate change related energy consumption needs to come down between 60% and 90% in the next 10 or at most 15 years. (see e.g. Hillman, 2004; Houghton, 2004; Henson, 2006; Monbiot, 2006). There are many creative and innovative approaches to this and huge efficiency gains can certainly be made. However, even the most optimistic assessments of what can be achieved in this way do not allow us to roll out the current conception of what it means to lead a "developed" lifestyle across a population of 6 billion people.

Moreover, there are deeper issues here. Technofix "solutions" can perpetuate rather than challenge the basic problem of over-consumption and hence simply cause worse problems further down the line in a way that Ehrenfeld, for example, calls "death by friendly fire" (Ehrenfeld, 2006). And technofix "solutions" perpetuate rather than challenge the underlying worldviews that many believe to be at the very root of our environmental problems.

Infinite earth and frontier ethics

What are these worldviews, and how should they be changed? First, modern industrialized societies are sometimes inclined to see the earth as a "sort of gigantic production system, capable of producing ever-increasing outputs" (HRH The Prince of Wales, 2007), or as a vast repository of resources. And, crucially, the way industrialized societies act appears in many respects to be underpinned by the assumption that this earthly production system or repository is infinite – both in terms of its capacity to provide us with resources and its capacity to absorb the pollution that our consumption of resources produces. This underlying assumption means that industrialized societies – and some forms of farming, astonishingly, have to be included in this – tend to operate in ways that ignore fundamental and unavoidable truths about biophysical systems. Ray Anderson describes this as "the linear, take-make-waste industrial system, driven by fossil fuel derived energy" (Anderson, 2007), operating as if the environment has no limits. This kind of worldview or mindset naturally inclines us towards an environmental ethic that has been called the "frontier" ethic. Frontier ethics tell us to make the most of these resources. Grab as much as of them as you can, as fast as possible!

The first necessary change, then, is to acknowledge that the earth's biophysical systems are circular rather than linear, and that they do have limits. We cannot endlessly extract resources at one end and endlessly emit pollution at the other.

ENVIRONMENTAL ETHICS AND ANIMAL WELFARE: RE-FORGING A NECESSARY ALLIANCE

Many resources, such as coal and oil, are effectively finite. We can use them up. Then they are gone. Those that are renewable – trees, fish, water – can be reduced to finite resources through overuse. And biophysical systems have a finite capacity to absorb pollution, the by-products of our industrial processes. We have seen this quite spectacularly in relation to the hole in the ozone layer and we see it in more mundane but equally threatening ways in relation to the consistent accumulation of toxic wastes, to loss of water and air quality and, of course, to climate change. We cannot extract resources at one end and emit pollution at the other and expect the pollution to go away. It will continue through the system, with various kinds of consequences.

Shallow environmental ethics –and humans on the edge

Acknowledging and responding to these environmental realities takes us to what has been called "shallow" environmental ethics. Shallow environmental ethics tells us that resources are finite and that we really do need to look after them. It asks us to act within the earth's limits: to use finite resources carefully and not to overuse renewable ones. Shallow environmental ethics argues that we have responsibilities toward the environment and other living things; it understands these responsibilities to be those of carefully managing a suite of natural resources and natural systems in our own interests and those of future generations.

Constraining our activities in relation to the earth's limits is clearly an absolutely crucial first step. It is particularly important in relation to intergenerational equity – seeking to avoid a situation in which we deplete environmental resources now at the expense of people in the future. But shallow environmental ethics, important though it is, is not enough.

For one thing, it doesn't challenge another aspect of the problematic worldview implicit in industrialized societies. This is that humans are somehow on the edge of ecological systems. We are managing them for our own ends, from a slightly detached position. You might call this the allotment mindset. The environment is out there, and we go out and take from it when we need to. We have to look after it – but we are not really *in* it. The most extraordinary form of techno-optimism – the view that sooner or later we won't need to be so careful because we will find ways of manufacturing our own resources – sometimes accompanies this. So long, earth, and thanks for all the fish – but now we can make our own.

This of course is absurd. The second critical worldview or mindset change is to acknowledge that we have a much profounder relationship with natural systems than the need to marshal them as a set of resources from a position of detachment. We are not on the outside of ecological systems looking in. We are very much on the inside, quite literally part of ecological systems, not apart from them.

We are not separate but deeply embedded. And we cannot transcend ecological systems. Our experience of life may increasingly distance us from the source of all our basic needs in ecological systems. But however many layers of technical brilliance intervene between natural resources and our end products, technology cannot detach us from our ultimate dependence on ecology. For all our technology, we remain earthbound creatures, relying on ecological systems for our most basic needs, as the MEA report points out. And we are not just dependent but interdependent, in the same way that all living things exist interdependently within the vast, extraordinary dynamic, interrelated complexity that is "the environment" or life on earth.

Human-centered/earth-centered ethics

The third mindset problem is the view of earth and other living things as a vast store of resources for humans. This is at the heart of shallow environmental ethics. The ecologically informed version of this mindset recognizes biophysical systems as a source of resources and of other "ecological services" such as clean air and water. An ecologically sophisticated version of shallow environmental ethics could even just about take interdependence on board, accepting that we are part of ecological systems and that harm done to them will sooner or later rebound on ourselves. But the bottom line is the same. The value of other living things, of habitats, of ecological and biophysical systems is considered to be instrumental, and only instrumental. Any value they have exists only in relation to their usefulness, in various ways, to us: and this is our sole reason for caring about them.

Of course, the environment *is* a resource for humans, and like all species, we have to relate to it partly in this way. But it is not only a resource. The vast complex of astonishing diversity, energy, and sheer will to live that is "the environment" has value that goes far beyond its usefulness to us. Blue-tits and basking sharks, savannas and rainforests, clouds, stars and streams have value beyond the extent to which one species amongst millions happens to need them. To deny this is to take an astonishingly arrogant stance, positioning humans as the only species of true worth and the rest of relevance only in relation to ourselves. This is a pre-Copernican view of ethics; the human centered values equivalent of the belief that the sun revolves around earth.

The third critical mindset change is thus to acknowledge the intrinsic as well as the instrumental value of other living things and systems – and to act in a way that respects this value. We have other reasons for caring for life on earth beyond the indisputable truth that we need it. This is sometimes called "deep" environmental ethics though I think a better term may simply be "earth ethics." Earth ethics tells us that the ultimate source and measure of value is not ourselves, and certainly not our economic systems, but the bigger context of which we are a part – the earth itself on which we, and our economies, inescapably depend.

From problematic worldviews to the ethics of earth-as-community

Taken together, these problematic mindsets, the view of "the environment" as a source of infinite resources, whose value exists only in relation to our own needs and concerns and in relation to which we exist in a sort of detached and – in our aspirations – independent state, add up to a worldview that underpins industrialized societies' relationship with nature. This includes many industrialized farming practices. Now, there is no clear causal relationship between this worldview and our current environmental problems. But equally it is clear that this view of the world leaves us feeling entitled and able to exploit it, at best within limits, at worst, in the naïve belief that technology will allow us to transcend all limits eventually. In various ways, this worldview facilitates, supports and justifies industrialized societies and industrialized ways of meeting our needs that, in treating other living beings as commodities, have failed so spectacularly to find enduring ways of living alongside them.

If we are to turn the current situation around, this worldview is amongst the key things to be challenged and changed. We have to move from seeing ourselves as detached managers of a set of resources to seeing ourselves as profoundly embedded within ecological systems. We have to shift from shallow environmental ethics to something very much deeper. Exactly how this deeper earth ethics is articulated will depend on the metaphors we choose to express our revised view. Understanding the earth as a living organism – a great blue and silver bird, perhaps, flying and soaring through space – certainly helps make vivid the idea that we, as part of this organism – cells, say – depend utterly on it. Moreover, if we are cells within a bigger organism then what we do clearly has much potential to affect it, and therefore us. But I think the metaphor of community helps to articulate how we should behave and treat others in a way that humans-as-cells doesn't quite capture.

Understanding ourselves as members of a vast community of other living things and systems, whose value extends beyond their usefulness to us, implies an environmental ethic very different from the shallow ethic discussed above. "Earth-as-community" ethics tells us that the continued functioning of the community as a whole, and the well-being of individual members of it, are both important, and it asks us to act accordingly. A pragmatic sensitivity towards others in the face of interdependence *and* a deep respect for others in their own right are both implied by the community metaphor. Of course we all have to use each other in various ways. But other living things are to be treasured and celebrated in themselves as well as recognized as necessary for our own survival. A further point is that the sheer complexity of this community of interconnected, interdependent beings and systems implies practical as well as ethical limits on our aspirations to control the environment for our own ends.

All this is summed up beautifully in a recent essay by Robert MacFarlane. Writing in relation to climate change, Macfarlane also captures and distils the bigger picture:

> We need to revive an older form of human awareness, according to which nature is neither backdrop nor storehouse, neither product nor chattel, but a community of which we are inevitably a part. It has become a compelling need that we move towards a respectful cherishing of the human world, rather than aspiring towards its complete control. The American ecologist Aldo Leopold saw this fifty years ago, before climate change had even been detected. 'We abuse the land because we see it as a commodity that belongs to us,' he wrote. 'When we see land as a community to which we belong, we may begin to use it with love and respect.' (MacFarlane, 2006)

Desperate dead ends

More of the same only louder

It has been argued that further intensification of animal-based agriculture will be a necessary part of our response to climate change. This will not be good news from an animal welfare perspective. But the urgency of climate change, it is argued, means that animal welfare becomes an issue of marginal importance, a luxury we simply cannot afford in the climate change context.

Leaving aside the highly contested question of whether further intensification and the correlative continued high dependence of agriculture on fossil fuel based energy *will* actually enable us to reduce our carbon footprint, it should be clear from the above why this advice is so profoundly mistaken. The worldview and ethics identified above and fingered as contributing to environmental collapse are, of course, exactly the same as those already identified as underpinning the problematic treatment of animals in intensive farm systems. It is precisely this treatment of other animals and other living things purely as human resources, as things or products, which is at the heart of the problem. And, as Einstein said long ago, you can't fix a problem with the same kind of thinking that caused it. We have to face up to the need for significantly new thinking and new directions rather than stubbornly persist in the current trajectory. Dealing with climate change by bringing about more of the same can never work.

Intensifying agriculture as a response to climate change *cannot* be the solution. It is as if, having learned that a loud noise is making us deaf, we respond by turning the sound up even higher.

Disassociating environmental issues from animal welfare

The attempt to prioritize climate change at the expense of animal welfare is in fact symptomatic of a split between animal welfare and environmental issues that precedes the climate change scenario.

Concern about environmental issues and animal welfare issues is often pursued separately, by different groups of people working for different organizations. Campaigning groups, for example, are often focussed on either environmental issues such as habitat degradation and loss, species extinction, pollution of various kinds, climate change *or* animal welfare ones. They do not always have much familiarity with the issues on the other side. At times there has been downright hostility between the two kinds of movement. I have been to animal welfare conferences at which many delegates had barely heard of climate change. And I have been at environmental campaigns where concern with animal welfare is marginalized or dismissed as a luxury or an irrelevance.

This split is understandable in various ways. The practical campaigns are underpinned by very different philosophical positions which have, amongst other things, a different ethical focus. Animal welfarists are concerned with sentient animals, especially domesticated ones. In their view, all sentient beings are ethically significant and should be treated as such. Environmentalists are typically concerned with the well-being, not of individuals, but of habitats, landscapes, species, and ecosystems – of ecological entities of various kinds. Whether or not these entities are sentient is considered irrelevant. And they are especially concerned with natural or semi-natural entities rather than domesticated ones. In their view, these ecological entities rather than individuals should be the primary focus of our ethical concern – whether this be for shallow or for deeper reasons (Rawles, 1997).

A further factor is the respective relationships between the environmental movement, the animal welfare movement, and science. The environmental movement has very strong links with science. Concern with animal welfare, irrelevant from a conservation perspective, has been viewed with suspicion as presupposing subjective mental states in animals – imagine! – within (some) scientific communities. It has sometimes been dismissed as sentimental and anthropomorphic. Given that many of the different elements within the environmental movement draw heavily on science for their authority, being associated with animal welfare might, in the past at least, have been resisted for fear of a loss of credibility (Rawles, 1997).

Of course, the animal welfare movement has itself benefited greatly from scientific work, not least that carried out by Marian Dawkins, Temple Grandin, and others. This work has given scientific credence to the claim that animals suffer in certain kinds of husbandry systems and has yielded valuable information about the natural repertoires of different animals and which aspects of those repertoires are most important to them. This in turn has supported groundbreaking work at the Food Animal Initiative (FAI) and other farms in relation to redesigning husbandry systems that facilitate rather than suppress these important behaviors. Nevertheless, on the whole, the animal welfare movement is perhaps more strongly associated with philosophy and particularly on ethics, its authority rooted in Midgley (1983), Singer (1991), Rollin (1992), and others.

The split, then, can certainly be understood. And, to a certain extent, it has resulted in a sensible division of labor. But it has also meant that groups who could usefully have been allies, calling together for better treatment of the other-than-human-world, have been fighting alone or even positioned as adversaries. In the current situation it is imperative that the two movements work together rather than compete. Despite the surface differences outlined above, they share an underlying concern to tackle the root causes of animal welfare and environmental problems. These are not really separate issues but part of the same phenomenon – the instrumentalization of other living things. Seeing issues as separate when in fact they have shared roots is one of our systematic problems, and pursuing these causes in isolation is an example of the kind of failure of joined up thinking that we simply can't afford. Pursing environmental gains at the cost of animal welfare is not the way forward, it is a way of staying stuck.

For animal welfare, the division between animal welfare and environmental issues has, in at least one important context, been highly detrimental in a more specific way. It has contributed to the almost complete neglect of the whole area of animal welfare in one of the most influential concepts of our time: sustainable development.

Sustainable development

The idea that societies need to develop in ways that are sustainable is – in theory at least – almost universally endorsed. Since the concept first came to prominence at the 1992 United Nations' Conference on Environment and Development – better known as the Rio Conference, or Earth Summit – "sustainable development" has become a guiding policy principle, adopted by governments, businesses. and organizations across the world.

Sustainable development encompasses a range of different goals and has been defined in literally hundreds of ways. Most definitions, however, refer to three main kinds of goal: social justice, economic development, and environmental protection (or similar). This is often summarized in various versions of the "sustainability triangle" with one of these goals on each point.

It is this characterization that has lead to the systematic neglect of animal welfare within sustainable development. Animal welfare simply doesn't feature on any point of the triangle. Social justice typically refers to justice within human societies rather than justice across species. Economic development is pursued primarily as a means of enhancing *human* quality of life. The environmental protection corner would probably have been the best bet but, because of the environment/animal welfare split outlined above, animal welfare is not typically considered here either. As a result, animal welfare is all too often not included as part of the goals of a sustainable society or discussed in the context of sustainable development policy. And, as sustainable development gets more and more prominent as a guiding concept, animal welfare risks getting left out in the cold.

The omission of animal welfare is a major failing of the sustainable development approach. I have argued elsewhere (Rawles, 2006) that the sustainable development triangle should be replaced with a diamond, with animal welfare as the fourth point. Unfortunately, however, this would not be enough to make sustainable development, as it is predominantly understood, a solution rather than a dead end. Sustainability, it is argued in the introduction, should mean having it all – human health, animal welfare, a flourishing environment. In fact, the way in which the term sustainable development is most commonly used in mainstream contexts almost guarantees that we will, in the end, have none of these things.

There are two main reasons for this. First, while sustainable development does include concern for the environment within its remit, this concern is almost always rooted in the kind of environmental ethics that we have argued above is profoundly inadequate. Take, for example, the "Brundtland" definition of sustainable development as development that "meets the needs of current generations without preventing future generations from meeting their own needs" (World Commission of Environment and Development, 1987). This is probably the most widely known definition and certainly a highly influential one, on which many others, including that of the UK Government, are modeled. In it, the environment is not actually mentioned at all! The environment features only by implication – as a necessary means for meeting the needs of humans. Other living things and systems are valued only as resources or commodities. We are back to shallow environmental ethics and humans positioned outside the environment, managing its instrumental value.

The most fundamental problem in relation to sustainable development, however, is that it is often used as a euphemism for business as usual. It can be used – or misused – in a very uncritical way in relation to western industrialized paradigms of development. This is the second main reason why the mainstream understanding of sustainable development cannot help us. What mainstream sustainable development seeks to sustain is the industrialized lifestyle and worldview. And this, as we have already seen in relation to its massive environmental impacts, is unsustainable. It cannot be continued into the future and it cannot be shared across the world's human population. Of course, the idea that the way we live should not undermine our own resource base is an important one, and the concept of sustainable development has been hugely effective in alerting people to a wide range of environmental and social problems – and to the connections between them. Nevertheless, this concept cannot, in its mainstream sense, give us what we need to move forward. Without major overhaul, "sustainable development" is more of the same, and another dangerous dead end.

Big bold solutions

To conclude, then, animal welfare and environmental problems have a shared root cause in the mindset that sees others in purely instrumental terms as a set of resources for humans; and that sees ourselves as detached and separate managers of these systems. This worldview, and the inadequate shallow environmental ethic that accompanies it, are amongst the most significant things that need to be tackled if we are to respond to the clear wake-up calls that are coming from many quarters – from climate change, from the desperately accelerated extinction of our fellow species, and from systematically poor levels of animal welfare. These issues are all connected and cannot be tackled separately. To take them together is to see that industrialized societies are heading in the wrong direction and that profound changes are needed.

Farming is both implicated in this and strongly positioned to show the way forward. Farming affects all of these issues – animal welfare, the environment, human health and well-being. And farming and food production affects us all. We all have a stake in its future. What sort of farming with what sort of ethics, underpinned by what sort of worldview do we want? One that leads towards ecological disaster or one that leads us towards a saner, healthier, fairer future for all? The general answer is clear.

To get there, we need to re-forge the ancient contract between humans, animals, and the land, and understand ourselves as members of a living ecological community in which others are treated with respect. This does not mean treating them as sacrosanct and unusable but it does mean treating animals as sentient beings with social, behavioral and other needs, and it does mean working with the grain of living systems rather than against, ensuring that farming is compatible with biodiversity and minimizing its climate change impact.

ENVIRONMENTAL ETHICS AND ANIMAL WELFARE: RE-FORGING A NECESSARY ALLIANCE

What this means in practice is being worked out at FAI and other farms, and is explored in other chapters. One conclusion pointed to by many of these factors and by many analysts is that, overall, the world's farming needs to involve fewer animals, leading a higher quality of life. This apparently goes against consumer demand. We are told that consumers want cheaper and more meat. But we also know that this is not compatible with a sustainable future in any sense of that phrase. Consumers as citizens, as Bonney argues (Chapter 6), clearly do want there to be such a future. Sooner or later this will be translated into market demand. Farming is of course compelled by business imperatives but in addition it can and should demonstrate leadership here – ethical and sustainable leadership. It should promote agricultural systems based on respect for other forms of life because that is the right ethic as well because such respect is implied by the worldview we need to move towards, if we are to continue our tenancy on earth.

And farming can help us experience as well as know what this means. We can intellectualize ourselves into a better environmental ethic only so far. We need to feel it too. Modern ways of living leave us feeling increasingly disconnected from ecological systems and other forms of life. Anyone who has been involved in the husbandry of fulfilled animals on farms that co-exist with a rich diversity of wild species knows how truly and powerfully farming can reconnect us with meaningful, sustainable and ethical ways making our living on this, one, earth.

Part 2
Bringing about change

We saw in Part 1 that the pursuit of productivity and profit in animal farming can lead to animal abuse and to the neglect of the welfare of individual animals. But while the pursuit of profit without an ethic of good husbandry is clearly not in the interests of animal welfare, it is equally true that trying to farm ethically without taking into account the commercial viability of what is being proposed is also bad for animal welfare. Animal welfare needs commercial viability just as much as it needs ethics. Otherwise the farmers trying to farm ethically will not be able to make a living, animal products will be sourced from less ethical farms, and there will be no net benefit to animals at all. Animal welfare needs to be *aligned* with what makes business sense and there is no need to be suspicious of business decisions that help animals just because they are also commercially viable. In fact, we should be more suspicious of companies that do things that are good for animal welfare without being able to see what they can get out of it because they are the ones that are likely to go out of business.

If the ancient contract is to be rewritten for the modern world it will have to ensure that farmers can stay in business. "Business" may sound ruthless and money-driven but it actually means "making a living" and the farmers of the future have got to be able to make a living or they will cease to be farmers. So, in Part 2, we now turn to the commercial realities of animal farming in the future. We use current examples of profitable farming that has animal welfare as a core value to point the way to where farming could go in the future.

It may surprise many people to know that animal welfare is already widely seen as an important commercial goal and already part of the strategic thinking of both small and global businesses, as Bonney explains in Chapter 6. Simple and practical ways of defining what we mean by good welfare (Dawkins, Chapter 7, and Grandin, Chapter 9) have now been developed into profitable enterprises by pioneering farmers willing to take the risk of doing something new (Layton, Chapter 8) and then taken up by supermarkets (Waterman, Chapter 10). Straight-talking advice on how to improve welfare and profitability at the same time (Grandin, Chapter 9) has been adopted by giant corporations as a necessary part of what they do (Kenny, Chapter 11). Legislation and farm inspection schemes then become an important way of ensuring that improved welfare becomes – and stays –

the norm that is universally expected (Main, Chapter 12, and Spedding, Chapter 13). And in case anyone has the mistaken idea that animal welfare is a concern only in Europe and the US, Paranhos da Costa (Chapter 14) emphasizes that it is also an important business driver in other parts of the world. Browning (Chapter 15) argues that many important ideas for the future are already to be found in current organic farming practices and Tudge (Chapter 16) sets out a blueprint for a more ethical approach to agriculture. Taking the chapters in this section together, it becomes clear that efficient farming, animal welfare, and value-driven consumers are building a surprisingly profitable partnership. It is not easy and there is still a lot of work to do, but done in the right way, good husbandry pays.

6 The business of farm animal welfare

Roland Bonney *was a sheep farmer for many years and has considerable experience of practical agriculture. He helped to form the Food Animal Initiative and has been instrumental in building links with developing countries. He has been particularly involved with developing model farm projects in Brazil and China.*

"People want to feel good about what they already do," a senior communications executive once told me. This is true; nobody wants to feel bad about what they are doing, or even worse, to be told that they should feel bad! Yet when you show our urbanized meat, dairy and egg eaters some of the more intensive production systems, the vast majority are far from happy with what they see. Indeed, I have seen experienced, commercial food company managers react violently to witnessing, at first hand, a "conventional" intensive pig finishing unit – by denying that this is, in fact, how much of our pork is reared in the UK. It is interesting that very few of these people want to go vegetarian. What they wish for is to be able to buy products from farming systems that do not make them feel ashamed.

Farming for food is not just about feeding people. It is about feeding people well. By "well," I mean not only providing people with enough, affordable, nutritious food, but also providing them with a sense of *well-being* about how that food was produced. This is never more important than when sentient animals are involved in its production. A sense of well-being comes when people are able to consume food that is produced in a way that reflects their ethical aspirations with regard to the environment, fair trade, and animal welfare.

Diet has always provided a fundamental definition of civilization and human culture, with food revered in a way that reflected the effort expended to acquire it. In the distant past, nomadic hunter-gatherers created cave paintings to honor the animals that fed them, even today the Inuit still say prayers of thanks over the carcass of the prey they have successfully killed and in the nineteenth century poets wrote ballads about the livestock herders of Northumbria and Australia. Getting enough food was hard work, something the whole community was directly involved in, and they gave thanks to the animals providing them with what they needed to live.

Thus, good animal welfare for our farmed animals came first, not so much because we cared about them, but rather out of a respect for their value in providing for us. This is what some have called "The Ancient Contract": the "deal" was that we provided them with a constant food supply, protected them from predators, and cared for them when they were sick and injured, while they provided us with food, clothing, and other useful products. Today we are consistently breaking that contract. In our relentless push to drive food costs down the individual value of some animals has been reduced to zero. They are worthless. Bull calves are killed at birth because dairy cattle have been so heavily selected to produce milk that many of their purebred, male offspring can no longer be viably reared for beef. Similarly, male chicks from egg-laying chicken breeds are killed as soon as they hatch because they are unsuitable and uneconomic to rear for meat. The value of breeding sheep is so low that it is cheaper to kill them than treat them when sick. The cost of this system is also detrimental in other ways: chicken designed for meat yield struggle to walk as their young legs are unable to support their weight and we cut off the tails and teeth of pigs to stop them damaging each other from stress-induced biting, caused mainly by the way in which we house them. Can we really say that these are the acceptable costs of progress?

With the recent development of industrialized production, food has become much more plentiful, reliable and cheap in the developed nations of the world. Nowadays, in the UK, for instance, 1% of the population produces 60% of the food that 60 million people eat. This is an amazing biological fact – that so few people can feed so many – yet now we take it for granted. As a result, the vast majority of us are no longer involved, associated with, or even witness the process of food production.

There has evolved a true commoditization of many foodstuffs during the twentieth century, with the likes of wheat and pork bellies traded on international stock markets. Food has become an input into a system that tells us that we can have as much as we want, for less than we paid yesterday. Meat and grain is often simply referred to as "raw material" in an ever increasingly sophisticated and efficient supply chain. But all this may be about to change, as the drive to reduce our impact on the world's climate together with the rising human population puts pressure on food supplies; cheap food may become a thing of the past.

> Food prices are set for a period of "significant and long-lasting" inflation because of demand from China and India and the use of crops for biofuels, according to Peter Brabeck the head of Nestlé. (Financial Times, July 2007)

In Europe we have lived through the era of butter "mountains" and milk "lakes" as a result of a Common Agricultural Policy which was designed to deliver affordable, secure supplies of food. But food cannot be sustainably delivered at a cost of damage to the environment and should not be delivered at a cost to animal welfare. We are now both being forced to, and choosing to, take a greater interest in how food is produced and in the resulting impact on people, animals, and our environment.

The following explores our need to understand what our food animals' really need and want to support their welfare, some of the economic reasons why this has not occurred, and how the market and government supports, or fails to support, improvements to the situation.

What matters to animals

When we first start to think about other animals, other than humans that is, our initial approach, understandably, is anthropomorphic. We use the powers of empathy that we employ daily with other humans, because these skills are immediately available to us. This can take us part of the way towards recognition of an animal's needs. After all, many animals have similar nervous systems to us, they too bond with their newborn, they recognize friends and strangers, and so forth. However, it is important to recognize that other animals are not human and that they often have different needs that we do not recognize or identify with so easily. To the animals these needs can be of even greater importance than those we more readily associate with, and they can only be properly identified by observing their preferential behaviors more closely. In the past we had a better real-life understanding as many more of us lived and worked with animals on a daily basis. As agriculture has evolved to feed our detached urban communities this contact, and our inherent understanding, has been lost by the vast majority of us.

Farms were generally much smaller in the past, with fewer livestock, so it was easier to relate to the animals on an individual level. Our farms have grown in size and scale to the extent that two men can now be in daily charge of 160,000 chickens. It is impossible for them to be able to see each chicken as an individual and yet it is the individual's needs that we need to meet. This is a challenge, but one which we can overcome by developing systems that inherently recognize and support the needs of the individual animals by their very design. Such systems can be, and have been, created where animals are able to maintain their own welfare and where they are not constantly struggling against the odds, reliant on inadequate support mechanisms to do so.

A desirable system can be defined by its ability to support *good* welfare *consistently*, where regular or recurrent failures are not excused by aspirational standards. At present many of the predictions for positive outcomes are based on "ifs" – if you maintain the ventilation system, if you control the predators, if the weather is kind, if the staff are well trained, if the feed is of a consistent quality, if the alarms are set – the list often seems endless. These are the constant challenges the farmer faces on a day-to-day basis when trying to manage his operation effectively. The more "ifs" there are, the more likelihood of failure. Reducing the amount of the "ifs" by creating systems where the animals can manage their own well-being and needs must be the goal, particularly in a world where labor and infrastructure costs continue to rise.

There are two critical things we need to do. The first is to have a reliable understanding of the animals' needs, not just what *we think* is most important. This is where the science of animal welfare provides us with an ever-increasing body of evidence because we can actually "ask" the animals themselves. It is through the work of scientists such as Dawkins and Grandin that we are able to put the animal at the center of our decision-making process, when building and designing production systems that better support the animals' needs. Secondly, and equally important, we need systems that the average person, not just the exceptional person, can operate effectively under all circumstances and not just on a good day when the sun is shining. We need to be realistic as to what we can expect of people with limited resources available in a low margin industry.

This is the work we have embarked upon at the Food Animal Initiative, not only to deliver systems that support the animals' most important needs but also those that can be consistently and viably operated with the resources available, and in the market place of, today.

The economic model

In the modern world of food production and agriculture, the language used is more often akin to that of the book keeper or accountant – residual value, net margin, return on invested capital, feed conversion ratio – none of which reflects the sentience of the animals involved. As a senior food company executive said to me recently, it is important to get the decision-makers out of the boardroom to see for themselves the living reality their decisions support.

The economic model developed by the food and farming industry has, until recently, seen animal welfare and the environment as "external" issues. The era of CAP (Common Agriculture Policy) in Europe, and similar subsidy systems in the US, have defined the economic models that are still taught in every agricultural college today. These models are output based; they tend to show that the more we produce the more money we make. They are based on getting more for less: if we can get a few more liters of milk for the same costs we will make more money. Whole supply industries to agriculture have been set up to exploit this model, such as the animal breeding companies whose main selling points for their genotypes have been more output for the same input, achieved through highly complex and focused breeding programs that have selected for the most efficient animals.

Farming businesses benefited for a while by simply producing more but as soon as supply outstripped demand the price for the product dropped in the market place. While the consumer and the economy benefited from lower food costs, success for farmers became defined only by least cost production. In our enthusiasm for this model, and with an unremitting focus on production traits, we have bred animals who struggle to conceive, who struggle to walk, who cannot give birth unassisted, and who require almost 24-hour surveillance and support.

I have nicknamed this type of production the "Formula One Model." The analogy is that, although I can drive a car passably well and safely, if a Formula One racing car was dropped off at the farm and I was asked to drive it to the local village, there is a high likelihood that I would end up in the ditch. I do not have the sophisticated driving skills to handle a Formula One car. Yet this is comparable to some of the highly selected breeds that we attempt to farm today. If we take the modern, high yielding dairy cow as an example, extreme attention to management detail and high levels of infrastructural investment are required if they are to be farmed effectively and humanely. We cannot expect all farmers to turn into the equivalent of Formula One drivers.

However some agricultural production systems are now taking animal welfare and environmental requirements into account. In many countries in the world – from South Africa, to China, to the US and Brazil, as well as Europe – farmers are getting a price premium for meat, eggs, and milk produced to higher standards. But premiums only last for so long for most products in any market. Early adopters benefit, but as soon as supply outstrips demand the competitive market brings prices down again. In other words, premiums are only part of the solution.

Farmers are in many ways the victims of their own success. We have produced more food for less cost and this has helped to reduce food costs dramatically. In the UK, for example, more than 30% of average income was spent on food in the 1950s, only 8% today. At the same time farming incomes have suffered some of the lowest financial returns in living memory. The idea that a margin can be maintained by ever increased efficiencies, particularly in systems dealing with other living beings, is a myth. The drive for commodity production for plentiful supply has not ultimately benefited farming businesses. It has been achieved, particularly in Europe and North America, with substantial support from subsidy and preferential trade agreements but on a global basis at the expense of the environment, animal welfare, an increase in the risk of diseases, and farmer bankruptcies. If we continue to push for more of what we want for less, we will, quite simply, end up with less of what we need.

This may not paint a pleasant picture, but things are changing – and these changes represent new pressures on the agricultural model. Efficient, sustainable food production will become of mounting importance as the impact from climate change and rising demand from an increasing global population are felt. Food

THE BUSINESS OF FARM ANIMAL WELFARE

costs are already beginning to rise. If we are serious about feeding the world's growing population through a period of unstable climate, we need to invest now in a viable and socially acceptable agricultural industry to ensure reliable supplies of affordable, good quality food. We need to develop new approaches to our production methods, not least because many of our current agricultural systems are heavily reliant on fossil fuels, particularly for artificial fertilizer production and for fuel to plant, harvest and distribute crops. For this to happen farmers need a margin, otherwise the necessary investment cannot be made.

Where do we look for solutions and address the shortcomings of the economic model we have been operating under the last 50 years? How much of these expectations can be imposed on farmers? How do we support them? Who should also take a lead in this process? Is it consumers, government or industry? Let's first take a look at the dynamics in the market place.

The market place

To paraphrase George Bernard Shaw's famous quote, "Democracy is a device that ensures we shall be governed no better than we deserve" – the market is a device by which we shall be fed no better than we deserve!

Within Europe, and to a degree worldwide, there is a growing sense that the market can resolve consumers' aspirations for higher welfare food products. This, it is argued, could remove the need for further legislation that is often contentious, time-consuming, and difficult to get agreement on, particularly in common markets where the particular interests of many nations are represented. Such legislation is seen either as a tool to act as a trade barrier against foreign imports, or as detrimental to local producers, who are required to adhere to higher standards, while cheaper products can be imported into the market from countries that do not have the same requirements. The theory is that by pushing the decision-making back onto the market place these criticisms are overcome as the choices are made by those buying the products and not by government.

The problem is that the ethical aspirations of "citizens" are usually higher than those of "consumers." We all have a dual personality – that of consumer and of citizen. The consumer within us articulates the way things "are," the citizen the way we think things "should be" – in other words, present and future. Most of us, when shopping for food or clothes, are in a different frame of mind than when we are reading about animal welfare issues in the newspaper. In the shop we find often bargains irresistible. Many in the food industry cite this is as an explanation as to why it is impossible to drive real progress, and in many ways it is not an unreasonable point of view. Why would a farmer or food manufacturer wish to put in place systems or standards that reflect aspirations but are not rewarded by actual consumer behavior?

While it is true that aspirations often run ahead of action, a food industry which chooses to ignore them does so at its own peril. What we say as citizens indicates where we wish to be as consumers – it is a form of advanced market intelligence. An example of this is the free-range egg market in the UK, and increasingly in other parts of Europe. In the UK more money is spent on shell eggs produced from noncage systems than on eggs from battery cages. Free-range eggs are increasingly seen as the norm of good policy in the food chain.

I once asked an experienced agricultural manager in the egg industry, if he had predicted this scale of growth 15 years ago, when the issue first became known to the public. He said he would never have imagined it. This is largely because until then the ethics of food production had had very little impact on the food market and the key drivers had been food safety, availability, and of course, price. His business had been rewarded by growing more for less, safely. These days the food market appears to be increasingly like any other market. It is dynamic and changing – the food we choose to eat is part of how we see ourselves. In other words, when our basic need to be fed is met we add additional requirements.

Consumers buy on perceived value. This is not just based on price, but on a range of requirements balanced against price in a market place with extensive consumer choice. The most obvious requirement for a food product is that it is enjoyable to eat. If someone does not like the taste of lamb, for example, its price is irrelevant and purchase will not be made. Equally, if a foodstuff is perceived to be of better taste, nutrition or quality, such as freshly squeezed orange juice rather than juice from concentrate, the price paid can be in excess of twice as much. Ethical issues can also add to or detract from the perceived value. Veal, for example, acquired such a bad name in the UK in the 1980s and 1990s, because of the lobbying by animal welfare NGOs against the veal crate system, that it is now rejected by most consumers. On the other hand, a growing number of consumers have sought out organic products which offer a whole package of perceived benefits including better animal welfare, reduced environmental impact and reduced health risks despite a substantial price premium. At a recent meeting a commercial director from a leading UK supermarket forecast that the organic market could expand to 10% of all fresh produce and meat sales within five years. It is evident that consumers do not merely buy on price – they also buy on the basis of values that enhance their sense of well-being.

Values or benefits have fuelled the growth in organic sales, in locally sourced products, in functional foods, in differentiation within the supermarkets own brands (such as in the UK Tesco Finest, Sainsbury's Simply the Best, etc.), and in ethically produced logos such as Freedom Food and Fair Trade. In China, the growth in organic consumption is now the highest anywhere in the world – running at 80% currently year on year. The Chinese will also pay 50% more for locally produced traditional chicken, such as the black skinned chicken, because it is perceived to be a better quality product.

Let's return to the question of whether we should leave a fundamental ethical issue such as farm animal welfare to the market to resolve. People do care about the welfare of farm animals. A recent study by the European Union (Eurobarometer March 2007) showed that EU citizens gave animal welfare an 8 out of 10 ranking in terms of importance. Grown men in the UK broke down crying as the cows they bred and lived with for so long were slaughtered in the face of foot and mouth disease. When interviewed they talked about the loss of the cows, not just in terms of economics but in the sense of friends, some described them as almost part of the family.It is not just a European phenomena either. There may be less statistics to back it up with, but you can see it everywhere if you look. Farm workers in China pat their water buffalo, not merely because they are their source of income, but because they are sentient, responsive partners in everyday life who they care for. There is, among many who have spent time with these animals, a deep-seated sense of respect and kinship with them.

It is clear that people are very concerned when faced with the reality of poor animal welfare. However, they are not faced with this reality at the point of purchase. Free-range eggs have probably been such a success because the concept of being caged is something we humans easily relate to and reject for ourselves. It is easy for many to accept and understand how it is likely that chickens suffer in this situation too. Free-range is a simple message, which allows a product to be promoted in a positive way. If the sector had tried to sell free-range eggs as cage free I doubt that market growth would have been as great. It is easier to get people to "buy into a benefit" rather than "pay to net off a negative."

Policies that support a market-led response to the concerns of citizens can only be identified as successful if our consumer buying habits begin to reflect our aspirations as citizens. If they do not, then we should judge the initiative as unsuccessful – but it is the initiative which should be questioned, not the primary demand. At the moment there is evidence that consumers want transparency, more information, and better labeling on which to base their buying decisions. If this was provided but buying habits failed to change, many would claim this is evidence that people do not *really* care. This is not necessarily true, because while better labeling could be helpful it may not be enough to support a change in buying habits.

When buying food we often have other overwhelming considerations, not least time! To read every product to identify where it sits in terms of welfare, health (fat, salt, sugar, and others), sell by date, price, etc. is simply not a realistic expectation. Equally, easily recognized logos or star rating systems will not readily convey real values or benefits to consumers as they will still need to get the information that defines the logo from elsewhere. All this requires effort which may prove to be too big an ask of time-strung people, and without this effort any additional labeling information will not be meaningful. Therefore, mere labeling is not likely to provide us with a silver bullet to address the issue of animal welfare.

A substantial "driver" in the market place currently is one of brand trust and corporate social responsibility (CSR). Leading companies in the high street globally are beginning to include animal welfare as a component of responsible business practice and are reflecting this in their supply chain standards and policies. Some, such as McDonald's UK who converted to free-range eggs 10 years ago, are leading the way on these issues not necessarily because their customers are demanding it when buying their products, but rather because they believe it is the right thing to do and it reflects the wishes of wider society. This is an example of the market responding to the aspirations of the people and it is important that such moves are recognized and supported.

There is another issue in relation to the food market – that of low income consumers. You sometimes hear from those in food retail that they have a responsibility to feed poor people. At one level this is laudable, we know that price is a critical issue for those on low incomes, but that does not mean that they have a lower ethical concern. The challenge for those food companies that wish to address consumers' ethical requirements is to make ethically produced food available and affordable to all. None of us condone human slavery as a means of lowering production costs to support those consumers on lower incomes, why therefore would we condone poor animal welfare? Responsible food brands accept this and show their leadership through driving progress on these issues on behalf of *all* their customers. Companies that take this approach in the future will generate greater brand trust as consumers are able to recognize brands which believe ethical values are truly important.

Another issue with "cheap" food is that we take it for granted, grumble when the cost goes up but throw masses of it away. It is a bit difficult to argue that food is expensive when we as consumers waste so much. According to a recent study from the University of Arizona in Tucson households waste, on average, 14% of their food purchases. The centralized supply chains that deliver our food to us have invested heavily in keeping their waste figures as low as possible but of course have left how much we waste to us – it is not their responsibility and anyway the more we waste the more we have to buy from them! This issue of waste in the whole food chain leaves consumers looking pretty irresponsible.

It is evident that improvements to farm animal welfare are the responsibility of all involved in the food chain. We need farmers to recognize that they are not just commodity producers but should be able to take pride, not only in what they produce, but also in the way in which it is produced. For that to be possible they need to understand the real needs of their animals. This is where the science of animal welfare can really help, by enabling evidence-based decision-making in the design and management of animal production systems.

But good production needs support from the market place to be sustainable and food retailers can take active responsibility for the decisions they make on

THE BUSINESS OF FARM ANIMAL WELFARE

behalf of their customers; not just in relation to their purchasing habits today but also in relation to their aspirations and expectations for tomorrow. Equally government must deliver a legally enforced trading environment that is transparent; a transparency defined not merely by access to information but one where advertising and labeling conveys a clear and accurate understanding to the consumer.

And so finally to consumers – that group that no one (government or retailer alike) dares to openly direct but only wish to appear to serve – us the public. As consumers all, we need to take a greater responsibility to close the gap between our purchasing actions and our aspirations if we are to deliver up what we really want.

As with so many things in life it is not just a question of what we do that matters, it is also the way we do it. For the vast majority of us there is nothing wrong with eating a steak, a milk shake, or an egg, provided that the way it is produced does not meter out damage to the environment or poor welfare for the animals. There is a great deal still to do before good farm animal welfare becomes an permanent part of our food chain. We must continue to address the issues at stake, drive progress, and support a food and farming industry which farmers and retailers can be proud of and which does not shame consumers.

7   What is good welfare and how can we achieve it?

Marian Stamp Dawkins *has a life-long interest in animal behavior and what it may be able to tell us about the minds of animals. She is currently Professor of Animal Behaviour in the Department of Zoology at the University of Oxford and a Fellow of Somerville College. Her books include* Animal Suffering: The Science of Animal Welfare *(1980) and* Through Our Eyes Only? The Search for Animal Consciousness *(1993). Her latest book* Observing Animal Behaviour *(2007) shows how observation can be used as an alternative to experiments in the study of animal behavior.*

There are two essential first steps in drawing up a new contract with farm animals. The first is to be clear about what we mean by "good welfare." The second is to find ways of farming that both provide good welfare and also enable farmers to make a sustainable living. The ancient contract might once have involved just two parties – humans and the animals they relied on directly for food, clothing, and a host of other products. But a renewed contract has to operate in a much more complex world, with most of us out of direct contact with farm animals and relying on other people to deliver the products we want. As consumers, we cannot demand high welfare for farm animals and then ignore the farmers who have to deliver it because, in the end, good animal welfare needs farmers who are able to make a living. In this chapter I want to show how adopting a simple but practical definition of animal welfare and then carrying out research *in context*, that is, within a commercial setting, can help to develop systems that deliver both high welfare to the animals and a reasonable living to farmers.

What is good welfare?

Although many words have been written about the exact meaning of "animal welfare" (e.g. Fraser & Broom, 1990; Broom & Johnson, 1993; Appleby & Hughes, 1997), and it is clearly important not to oversimplify a complex issue (Dawkins, 1980;

Mason & Mendl, 1993; Webster 1994), sometimes it is even more important to get to the heart of the matter of what good welfare is. Despite a wide variety of backgrounds and experiences and views about animals, it is my experience that what most people mean by good welfare is that animals are healthy and have lives in which they have most of the things they want. Conversely, what they mean by poor welfare is that animals are unhealthy (diseased or injured) and/or they do not have what they want, where "not having what they want" could include both being deprived of something important in their lives and also being unable to escape from or avoid situations they disliked.

Saying that animal welfare is about animals being healthy and having what they want sounds like a gross over-simplification of a problem that should really be treated in much more complex ways. But having tried it out on a variety of people over the last few years and used it myself as a research tool, I have come to the conclusion that it is an extremely powerful way of approaching farm animal welfare.

Its first overwhelming advantage is that everyone can understand what it means and everyone I have so far suggested it to seems to agree with it. Only last week a poultry producer from a large commercial company challenged me to define animal welfare, obviously expecting me to give a small lecture on the subject. "Animals that are healthy and have what they want." I said. There was a pause. "I'll buy that," he said.

I've had similar responses from people coming to farming issues from quite different starting points. People already committed to animal welfare and even people who know little or nothing about animals or practical farming can all see the point of it and agree that it encapsulates what they mean by "good welfare." So let me explain a little more fully what it involves and how it, quite subtly, incorporates much more than appears at first sight.

The "health" side of welfare is usually uncontroversial. Most people would agree that, as an absolute minimum, good welfare means that the animals must be healthy, that is, they should not be dying of disease or slipping over and injuring themselves. But most people also feel that there is more to good welfare than just not dying. Bright eyes and healthy fur and feathers are a good start but there must be something extra for really good welfare. Of course there is. That "extra" is the second part of the definition: what the animals themselves want.

Good welfare means that the animals are content because they are not desperately searching for something they do not have and, equally, that they are not fearful of something they are trying to get away from and can't. "Having what they want" thus includes having the positive elements needed to satisfy them and *not* having to put up with the fear-producing, anxiety-producing, boredom-producing elements that they want to escape from or avoid. The same idea could be expressed by saying that animals should as far as possible have positive emotions (the ones we call pleasure or contentment) rather than negative ones (the ones we call fear, pain, boredom, hunger, exhaustion, and so on). The advantage of talking about animals "having what they want" rather than having "positive emotions" is that it is much clearer what we need to find out about them. Different species express their emotions in different ways but we can set about the task of asking the animals what they want and what they don't want in clearly defined behavioral tasks such as offering them choices or seeing if they will learn to avoid or approach things. We can let them tell us what is desirable or undesirable.

Laying hens, for examples, will work hard (learn to push a heavily weighted door) to gain access to somewhere to dustbathe and somewhere to perch (Olsson et al., 2002). Mink will similarly push weighted doors to be able to enter a swimming bath (Mason et al., 2001). Such findings indicate what features of an environment are important to the animals themselves – in other words, what they want. We don't have to worry about defining the exact emotional state of a mink without access to water (boredom? frustration?) or a chicken without access to a dustbath (itchiness? anxiety?). We just let the animals tell us what *they* want, whether or not it has any connection to what we humans would consider desirable or undesirable. We learn about what is desirable or undesirable from the animal's point of view.

The second reason why "healthy and having what they want" is such a powerful definition of welfare is that it helps us to make sense of all the other different measures of welfare that are now available, such as body temperature, increase in "stress" hormones, stereotyped behavior, the extent to which behavior is "natural," choice and preference, scales of health, and walking ability. There are now so many different ways of measuring welfare that the problem is not having enough measures but having too many and finding that they often give contradictory results. For example, laying hens prefer an enriched environment with grass and scratching areas to a barren environment with a wire floor, but the corticosteroid (sometimes called "stress" hormone) levels are higher in birds given access to the preferred environment (Dawkins et al., 2004). The "healthy and having what they want" definition of welfare comes to the rescue and helps us to make sense of what might otherwise be seen as a confusing result. The fact that the hens prefer the environment in which their corticosteroid levels are higher suggests that an increase in these hormone levels may not be an indication of negative emotions or undesirable stress at all. Rather, raised levels may be indicative of activity, or arousal and even associated with positive emotions.

In practice, it is often difficult to say what "raised levels of corticosteroid hormones" actually tell us about animal welfare (Barnett & Hemsworth, 1990; Rushen, 1991). Although corticosteroids are frequently referred to as "stress" hormones (an unfortunate term that prejudges their function), they also increase dramatically with experiences that humans find pleasurable such as sex and the anticipation of food (Toates, 1995). So-called stress hormones are therefore more properly associated with arousal or activity and what they actually tell us about welfare needs to be validated by relating them to either health or what the animals show that they want. For example, if high levels of corticosteroid predicted serious health problems or were consistently associated with situations that animals chose to avoid, then we could have confidence that they were a good "measure" of welfare. On the other hand, if they indicated excitement and activity and were equally associated with situations the animals consistently chose for themselves, then they would be seen as not saying very much about whether welfare was good or bad. Its what the animals themselves want, together with information about what makes them healthy or unhealthy, that are the best guides to their welfare.

"Being healthy and having want they want" is thus not an alternative to other measures of welfare. It actually includes them and makes sense of them. Another example of its usefulness is the controversy over whether it is essential for an animal's welfare for it to be able to perform all its natural patterns of behavior. On the one hand, freedom to perform natural behavior is one of the Five Freedoms that the Farm Animal Welfare Council (www.fawc.org.uk/freedoms.htm) proposed as the basis for good welfare. But, on the other hand, does this mean all the behavior in the animal's repertoire? Being pursued by a predator is natural for many animals, which would suggest that being hunted should be included in the requirements for good welfare? Can it really be good for welfare to be hunted? If not, is "natural" behavior useless as a criterion of welfare?

"Healthy and having what they want" clarifies the situation by making it clear that the two things we need to know are: Does making provision for the animal to perform the behavior improve its health and is it something the animal wants to do? For some natural behavior the answers will be yes and for some other behavior, such as being chased by a predator, it will probably be no. It isn't the naturalness of the behavior that tells us about welfare. It's the effect that behavior has on health and the animal's view of what it wants to do that are the deciding factors.

So "healthy and having what they want" does not replace all these other measures of welfare. It makes sense of them. It gives them "valence" and tells us what we really wanted to know all along – that is, whether the animal is in a negative emotional state (such as fear, boredom, frustration) that it dislikes and wants to get out of or a positive emotional state (pleasure) that it likes and wants to repeat or stay in. But the third reason why this is a good measure of welfare is perhaps the most important of all. This is that it clears our brains and tells us exactly what we have to do to improve animal welfare.

Quite simply, we should be farming in ways that keeps animals healthy and give them the conditions that they themselves want. Putting it like that makes it clear that a definition of good welfare that seemed so mild that most people could agree to it without much difficulty has real teeth. It does not refer to what well-meaning people think would make animals healthier or what they think animals want or even what they themselves would want. It demands a much more animal-centered and evidence based approach than that. It requires all the resources of animal welfare science to provide evidence about what actually does make animals healthy and what they really do want. It commits us to finding out, from the animals' point of view, what their worlds are like and what they need. But it also goes much further than this because it carries a commitment to go beyond just the scientific findings and the research papers. It means putting ideas into practice and making sure that when we try them out, the animals really would be healthy and have what they want. We need to know whether the farming systems we devise work as we require them to. That's what has to go into the contract. Not just fine words and armchair ideas, but practical solutions that deliver what all parties to the contract want.

Of course, animals may not be able to have exactly what they want, any more than the rest of us can, or any more than wild animals can, but we need to start with their wish-lists. Then we need to find out what items on those list are essential to their health and well-being as we take the next step of developing ways of practical farming that incorporate these essential elements of good welfare. This is not as easy as it sounds. A commitment to improving animal welfare, however, is about going global and making changes in the real world of commercial food production. The new contract with farm animals is not just about the future of a few small scale niche farms. It is about our contract with the billions of farm animals that provide humans with food so that contract has to deliver workable systems that give those billions reasonable lives. And to be deliverable and sustainable, those systems must protect the environment, deliver healthy safe food, and allow farmers to make a living. It is a tall order but it is an urgent one. The goal is no less than major changes in the ways in which animals are farmed across the world and we should be judging the success of what is achieved against two simple yardsticks: whether animals are healthy and whether they are leading lives that give them what they want.

'Rollout"

Governments and funding bodies the world over have given surprisingly little attention to the process of how research findings in any field can best be made use of by the "end-users" or "stakeholders" (in this case, farmers). The key is often assumed to be "knowledge transfer." This is usually seen as a process of explaining the results of research to convince "the public" or "industry" that the research has practical value and so should be taken up and widely used commercially, a process rather inelegantly referred to as "rollout." The implication is that if this is not achieved, there has either been a failure of communication or the research is useless or farmers cannot recognize a good thing when they see it.

Particularly in the case of farming, it is now clear that the process of rollout is a great deal more complicated than this. It is not enough just to communicate results of scientific research (although good communication always helps) or even to demonstrate a new or improved system working in practice. A farmer, living on a low income and only just making ends meet, is not going to risk his home and his whole livelihood on the off-chance that something demonstrated on a model farm or on a research establishment just might work for him or her. The agricultural "industry" (which may in practice consist of a collection of small farmers) is not like the pharmaceutical industry where large profits can be used to invest in long term projects such as developing a drug that may not come onto the market for several years. It may be true in some industries that if an idea is any good it will be "taken up" by the industry. But that is not true with farming. The lack of funds for developing ideas into commercially viable agricultural systems means that much research, even if it is "near market," remains just that – near market but never actual market. David Wood-Gush, to whom this book is jointly dedicated, spent years developing his "pig park" system in Edinburgh (Stolba & Wood-Gush, 1984, 1989). Having studied the behavior of pigs allowed to run free on a Scottish hillside, he devised a system of keeping pigs that incorporated as much natural behavior as possible. Family groups of pigs were kept together so that the social behavior was as natural as possible. The system worked very well and the work was supported for many years by the RSPCA. But no one ever took it up commercially. It remained a near market, near miss that got nowhere despite the obvious welfare benefits to the pigs.

Looking back, it was obvious why it wasn't used commercially. For a start, it was extremely complicated. There were bedrooms with straw for sleeping, dunging areas, feeding areas, and a complicated system of gates with different sized corridors to allow the smaller pigs to escape from the large ones. No commercial farmer would have had the time or the patience to run the system as it was and certainly not the money to develop the system so that it was easier and more convenient to run. This illustrates one of the many hurdles in moving from small scale prototype to having a real impact on the way animals are farmed commercially. It is the problem of development. A system may work reasonably well on a small scale but before it has any chance of "rollout" it needs a great deal of modification, trying out a new version, tweaking and debugging until it is ready to be used by a farmer who has to be able to make a living out of it. Yet small differences can make all the difference between success and failure. Working out how to get the best out of a system needs time, money, and resources, but the trouble is that it does not attract money. Development of a system is not "world class science" and so is of little interest to the research funding bodies. At the same time, if it is near market but not really near market enough, it will not attract industry money either, particularly from a hard pressed industry such as farming. To achieve roll-out on a large scale of farming systems that incorporate high welfare provision for the animals, we need a complete rethink of the way in which welfare innovations are put into practice. To fulfill our side of the contract with farm animals by providing practical systems that deliver good welfare, we cannot change the way individual farmers keep their animals by some ill-defined process of diffusion or "market forces." Rollout needs to be planned for, pushed and developed all within the framework of farmers having to make a living. Much of the rest of this book is devoted to the ways in which we can actually bring about changes in farming. We have to make things happen. They will not happen on their own, but a surprising number of people are now willing to be part of a joint effort to bring about change.

Making things work

Into the gap between ideas and practice, between prototype and commercial reality has stepped an extraordinary organization, the Food Animal Inititaive (FAI), based at Wytham, near Oxford. It is extraordinary because, on the one hand, it has core sponsorship from two of the largest food companies in the world, Tescos and McDonalds, and on the other, it works closely with Compassion in World Farming, the World Society for the Protection of Animals, and other animal welfare organizations. At the same time, FAI has strong research links with Oxford, Warwick and Bristol Universities. As if this weren't enough, they are commercial farmers and run their farm as a stand-alone commercial business (the core sponsorship does not go into the farm). The fact that they are commercial farmers is one of their most important attributes. They suffer the same problems of difficulties in getting staff, sudden changes of mind from supermarkets on what they want, ups and down in the market as all other farmers. Throughout this book, there are numerous references to FAI, not out of some anglo-centric view of the future of farming or because FAI is the only place where new approaches are being adopted (encouragingly, there are more and more like-minded farmers), but because FAI have so directly addressed the issues of linking research with practical farming and global business. FAI has branches in Brazil and China and it is regularly visited by American businesses and charities because they see it as a harbinger of a future of welfare farming that is already happening in Europe and will soon spread to the US. Its successes and even more its failures go far beyond its 750 hectares in the center of England and reach out to farmers trying to make a living everywhere.

One of the driving forces behind the setting up of FAI was to directly address the issue of how to make changes in farming so that there would be improvements in animal welfare worldwide. FAI provides a new model for making change happen. The ingredients of the new model are: close links between research and the needs of the farming community (by doing research in a commercial context in the first place), time and effort spent developing farming systems so that they work in a practical and commercial sense and genuinely deliver good animal welfare and, finally, rollout to the wider world so that the new systems are adopted not just by the few but by the many. The research is done through collaboration with universities, the development is done by painstaking trial and error on their part, all within the constraints of a commercial farm. The all-important rollout is done through their unparalleled links with the farming community, animal charities, and big business. In the next chapter, Ruth Layton describes how FAI took David Wood-Gush's ideas for a high welfare pig system that had remained on the shelf untouched for many years and developed them into a simple to run, highly competitive commercial system that large food companies are now pushing their suppliers to adopt. The new model is being put to the test and the results have lessons for anyone wanting to make a different future for animal farming.

To illustrate what it means to do research in a commercial context and also how it is possible to integrate welfare, human health, ecological impact with commercial success, we can use the example of one of the first research projects set up at FAI, the so-called PINE project (Poultry In Natural Environments). The aim of this project was to develop a commercially viable free-range system for keeping broiler (meat) chickens that could be easily adopted in small stages by anyone wanting to set up a systems with small initial cost. At the same time we wanted to be sure that the birds were not having an adverse impact on the environment and that the chickens would not be a human health hazard through their contact with wild birds. We also wanted to be able to design the best possible range areas so that the birds would see them as desirable environments and range widely.

The majority of broiler chickens in the world are kept inside in intensive houses but an increasing number are kept as organic or free-range birds. Many farmers are now beginning to enrich the range areas by providing their birds with tree cover. What was not known was whether this actually improved the birds' welfare or just satisfied our human ideas of what chickens "ought" to be doing. We also had no idea whether the birds would damage the environment (e.g. through nitrogen run-off) or whether there would be adverse disease implications of having large numbers of chickens outside. Biosecurity is, after all, quite impossible for free-range birds.

We combined a fully commercial venture designed to be profitable with a properly replicated experiment designed to be scientific (a combination that various people told us before we started was quite impossible to achieve). The project had to be profitable right from the outset because the only funding we had was for the research side. It also had to be statistically robust to satisfy the research funders. There were 16 free-range units, half planted with trees and half without trees so that we could compare the effect of trees on various aspects of chicken welfare, productivity and behavior. The chickens were housed in small moveable houses ("arks"), each holding 670 chickens. A steady stream of 1300 birds a week was supplied to a major retailer and over the course of the first 18 months of the study, we studied a total of 112 flocks, 56 of them having access to trees and 56 of them having no trees in their ranges, a degree of replication enough to satisfy even a demanding statistician.

The results were very encouraging. The system was a success commercially. Right from the start, the birds were marketed as a valuable supermarket commodity and consumers were prepared to pay more for the high welfare specifications. By the second year of the project, when the trees had grown considerably, plots with trees attracted more birds outside than plot with no trees, particularly on sunny days. More birds ranged outside in the summer and autumn than in winter and spring, correlated with external temperature (Jones et al., 2007).

Encouragingly, there were neither environmental or disease problems. Most nutrient levels in surface and groundwater were low before the chickens were introduced and remained low afterwards and all flocks were *Salmonella* free. One of the most fascinating results to emerge from the study was that there was no evidence that the chickens were becoming infected with *Campylobacter* through their possible contact with the wild birds around them. *Campylobacter* are, worldwide, one of the major sources of human food poisoning and eating contaminated chicken meat is one of the main ways people become infected. The strict biosecurity that can be enforced in intensive systems to prevent chickens picking up *Campylobacter* is, of course, impossible in free-range and organic systems and so there is always the worry that while ranging outside might be good for the welfare of chickens, it might also be a threat to human health. Martin Maiden and his group from the Department of Zoology in Oxford used the latest genetic techniques to "type" *Campylobacter* and showed that this was most unlikely. The *Campylobacter* carried by the free-range chickens were quite different those from wild birds found nearby, such as geese and starlings. Furthermore, the chicken flocks that ranged best (more birds came outside) had no greater risk of carrying *Campylobacter* than those where more birds stayed inside (Colles et al., submitted). The free-range environment thus seemed not to pose the health risk that people had feared.

This conclusion is obviously a considerable relief to free-range and organic farmers and to the people who buy their produce. The important point is that the conclusion has credibility precisely because the research was done in context – on a working farm, with large numbers of birds being reared and moved each week.

WHAT IS GOOD WELFARE AND HOW CAN WE ACHIEVE IT?

If it had been done on a small scale with just a few birds, people could still have argued that wild birds could be major source of infection. At the same time, if it had been done without the proper experimental design and measurement that were going on in parallel, the conclusions would not have been believed either. Doing the research in the context of a commercial farm is a basis of the FAI model and answers both sets of critics at the same time.

For as long as there are things we don't know or don't understand fully we will need research to find the answers. It may surprise some people to know that there is a particular need for research and development in nonintensive systems of agriculture because we don't always know how get the best out of them. We need the right diets, the right breeds, and the right management to ensure the welfare of the animals. It is also an unfortunate fact that there are often unexpected problems with what seem to be improved systems. For example, the EU decision to ban battery cages for laying hens has raised serious concerns for the welfare of the birds because of increased problems with feather-pecking and cannibalism in noncaged laying hens (Savory, 2004). This does not mean that we should continue to keep hens in battery cages, as there are other positive welfare benefits to noncage systems, but it does mean that we have to do some more work – research and development – to find how to keep hens outside cages in ways that they do not damage each other. It's a challenge to improve welfare, not a reason for giving up.

Defining good welfare in a straightforward way that everyone can sign up to is an important first step in moving towards higher welfare farming in the future, but in many ways it is the easy part. We know what it means to talk about animals being healthy and having what they want and we can identify the sorts of *in context* evidence that we are going to need. Putting good welfare into practice and making sure that under commercial conditions the welfare of the animals is ensured is much, much harder. But if the new contract is to mean anything at all, it must have clauses that guarantee delivery, not just promises. Delivery is the next step.

8 Animal needs and commercial needs

Ruth Layton *is a founder Director of the Food Animal Initiative (FAI) (www. faifarms.co.uk). She trained as a veterinary surgeon and then took the Certificate and Diploma in Animal Welfare Science, Ethics and Law at the Royal Veterinary College. In 1997, she started her own animal welfare consultancy, RL Consultancy, which brought her into contact with large scale producers and food retailers. The experience of seeing how farm animals are reared made her determined to try out alternative ways of farming, which eventually led to the founding of FAI and its move to Wytham, near Oxford, in 2001.*

Having a clear idea of how animals should be farmed is one thing. Putting those ideas into practice so that they work in a competitive commercial world is quite another. I realized this in a big way when, after completing the Royal College of Veterinary Surgeons certificate and diploma in animal welfare in the mid 1990s, I found work as a consultant working with large agricultural companies supplying a major food retailer. My naïvety at the time lead to considerable frustration on my part that my "pearls of wisdom" were not immediately grasped and acted upon by all companies with the same degree of enthusiasm with which they were given. Looking back, the reasons for this should have been obvious. Putting animal welfare into practice was never going to be just a question of making suggestions and then expecting others to immediately change their way of farming. For both large companies and smaller individual farmers there are many forces acting against change, including some of mind-set, some of commercial reality, and some simply because that's the way they have always done it. Change is hard work.

It became increasingly obvious to me that if we were to make real progress in farm animal welfare, we needed to take the whole process of bringing about change in agriculture much more seriously than had been done so far. In some areas it might be possible to make small incremental changes to existing systems to avoid welfare problems such as mastitis in dairy cows, predation in organic and free-range poultry systems, and parasites in sheep. In other areas, however, we might need more of a revolution than evolution to make any real difference.

It would not be enough just to tinker with an existing system. In other words sometimes evolution would have to be replaced by revolution.

It was at this point that I joined forces with Malcolm Pye and Roland Bonney to form the Food Animal Initiative (FAI) with the strapline: "animals fit for their environments – environments fit for animals" to emphasize that if we farm in ways that suit animals, people also benefit. Shortly after we took over the tenancy of the Oxford University Farm at Wytham a headline in the local press, "Animals design their own housing" picked up this idea perfectly. Below this headline was an article which described the work of FAI which encapsulated exactly what we were trying to achieve.

We had another strapline as well, specifically aimed at farmers: "We fall over so that you don't have to." We knew that in order to bring about real change in farming, we were going to have to take risks ourselves and try out new systems that other farmers could not afford. We recognized that to address the fundamental issues of agriculture we were going to have to invest time and energy in acquiring knowledge, putting it into practice, and making it work. This was going to be a long, complicated and inevitably expensive process. In addition, we would promote the success of those farmers who had already achieved break through in the way they farmed.

In this chapter, I will endeavor to describe this process of bringing about change and what we have learnt from our experience at FAI since 2001. We have attempted to tackle both evolutionary (step by step change within existing systems) and revolutionary change (new systems). While there is much we can do to improve welfare within existing systems, we also have to face the unfortunate fact that if we want to make real change, it is not always possible to "manage" our way out of the problems causing poor welfare. Openness to both types of change will help us as we think of changing the animals' environments to improve their welfare. We also need to ask ourselves whether we have the right breeds of animals to start with.

I would like to remind readers that the changes we have seen in agriculture over the last 60 years and which have had such a negative global impact on animal welfare were made originally in good faith. The main objective was to feed the nation after two world wars, when we did not have food stability and the population was on the increase. Moreover, we knew and thought much less about what was important to animals at that time as our need was so great. This is no longer the case. The problem before us now is how to meet animal needs and commercial needs, when the two often seem so incompatible.

Working with genetics

Many modern breeds of farm animals have inherent welfare problems because they have been bred for highly productive traits at the expense of their own welfare. That is why we have ended up with breeds of cattle that require frequent caesarean section as they cannot calve easily on their own, turkeys that cannot mate naturally as the male is so large and cumbersome, and breeding pigs and chickens that have to be restricted in the amount of feed they are given as otherwise they become too large or fat to breed effectively. This is not because they have been genetically modified by new high-tech methods for manipulating genes but because they have been selected by the same classical breeding methods that gave us domesticated breeds of dogs and all our traditional breeds of farm animals. Over recent years the breeding programs have become more and more intense and have concentrated on production traits – that is, on producing animals that give the most milk or the most meat and the highest growth rate for the least amount of feed.

If genetic selection has produced animal welfare problems, it can also be the solution, provided that we promote the selection of good animal welfare traits alongside production traits. If a new breed has both high welfare and good production, then it will make life easier for farmers and can be incorporated into existing farming systems with little or no disruption and indeed with advantages all round. Sometimes these breeds are already available and all that is needed is to demonstrate to farmers that the breed is useful and that they can immediately take advantage of it. This is one of the functions that FAI has undertaken. At other times, it may be necessary to develop a new breed of animal altogether because there is nothing currently available that fulfills both market and welfare needs. At FAI, we therefore see another of our functions as promoting the selection of animal welfare traits alongside production traits as a relatively new concept in the genetic development of farm animals. Our aim is that new breeds, with genetic selection for a combination of high welfare and commercially competitive production, become a major way of promoting animal welfare on a commercial scale.

Low maintenance sheep

About 40 years ago, some of the breeds of sheep native to the UK were taken to New Zealand and crossed to produce new breeds which do particularly well on the extensive grazing systems prevalent in that country. The most successful animals were kept for the next breeding generation and the New Zealanders created several breeds which are good foragers (actively search for and make best use of herbage) and thus can do very well on grass alone without the need for significant quantities of increasingly expensive grains such as barley. They can give birth without problems, have low lamb losses, and are hardy in terms of their susceptibility to disease.

Now that the financial margins on sheep production have become tighter, with increases in labor and feed costs, many lowland sheep farmers in the UK are changing the way they manage their sheep. Instead of bringing our sheep inside at lambing time, which is relatively expensive in terms of labor and equipment, many of us are turning to outdoor lambing systems. For this to be successful, we need a ewe that can perform well in such a system and, with minimal human intervention, give birth and rear her lambs. A sheep that can take care of herself and perform commercially is invaluable to a struggling sheep industry and could be highly beneficial in welfare terms when combined with effective management. At FAI we looked at the sheep the New Zealanders bred from the original stock and decided to buy rams from a breed known as the Coopworth which had been brought into the country by an innovative farmer in Scotland. These were used to mate our current flock of 800 traditional "mule" (crossbred) ewes which we bought each year.

To achieve the characteristics of the flock that we most value, we choose for breeding only the females that meet certain criteria. Instead of natural selection acting on a wild population we are the ones doing the choosing. We select for breeding only those ewes that are good mothers themselves and give birth without assistance. Any of our ewes that have trouble at lambing time are marked and not used for the next breeding program. Thus we select female lambs only from those ewes that have twins, have no problems at lambing time, and whose lambs grow well up to and after they are weaned from their mothers.

Our task is made much easier by having the Coopworth rams which already have many of the genes we are looking for and which contribute half the genes to the females from which we make our selection. There is the added advantage that the breeding females are brought up in the environment in which they themselves breed so that they develop background immunity to many of the diseases on the farm. This immunity they then pass on in turn to their offspring through colostrum (first milk produced at birth) each year. As the flock is now closed (no new ewes brought in), the females have an increasing percentage of these valuable Coopworth genes and there is a reduced risk of bringing new disease onto the farm. New Coopworth rams are bought on a regular basis to maintain genetic diversity, but only after a long period of quarantine. The principal behind this strategy of breeding from within the home flock in the environment the lambs will grow up in is that we end up with animals that do well in *our* system and that are best suited to be the next generation of breeding stock on *our* farm. This has disease and therefore commercial advantages.

The traditional way of managing lowland sheep flocks in the UK is to buy in new cross-bred females each year from the hardy hill sheep found in the northern part of the UK. Although this system has the advantage of hybrid vigor (the hardiness that comes when two animals of very different types are bred together as the "strengths" of each animal are carried through to the offspring), it carries an increased risk of spreading disease as the sheep are moved from one area to another. This system shows that there is a viable alternative for managing flocks of lowland sheep and farmers can decide for themselves in a changing market which best suits their needs.

Broiler (meat) chickens

A different situation arises when there are no existing breeds anywhere in the world which would combine production and welfare traits sufficiently well. Over recent years the relatively new discipline of animal welfare science has shown us that modern broiler chickens can grow so fast that their bone and cardiovascular development often cannot keep pace with the physiological needs of their bodies for bone strength and blood supply. A side effect of this selection is that the birds kept for breeding continue at a high growth rate into adulthood and often have to be kept very hungry in order to prevent their potentially excessive body weight resulting in problems such as obesity, lowered male fertility due to inability to mate successfully, and low female egg production.

There are thus major welfare issues in broiler chickens both with the juvenile (meat) stage and with the adult breeders. There are breeds of chickens that have no such problems but they are so slow growing and have such poor meat yield that they are uneconomical to produce. Research carried out by retailers in the UK has shown that even if the price were comparable to current chicken prices not enough customers would be interested if all they got was a bird with a narrow breast and large legs. Breast meat is preferred by most consumers in the western world. Trying to produce and sell such birds would have little impact on bird welfare or farming incomes if customers would not buy them. It might be possible to spend a lot of time and money trying to educate customers to change their buying habits but we believe that this is unlikely to work.

We feel a more effective solution is to produce a new breed of bird that has reasonable breast meat yield and yet does not have the attendant welfare problems. Even chickens kept in organic and free-range systems have some of the same problems that are seen in the breeds used in the more intensive systems and this is particularly so in the parent birds. These issues are frequently highlighted by the media and animal welfare organizations. Our aim is to produce a bird which has good welfare, meets the eating requirements and price expectations of customers, and allows farmers and others in the food chain to make a living. Believing this to be possible and to be a very important goal for improving the welfare of broiler chickens, FAI has just started a major new breeding project.

ANIMAL NEEDS AND COMMERCIAL NEEDS

Fundamental change in systems

Important though developing new breeds is for the future of animal welfare, there are some systems of agriculture that are not at all valid for farmers or consumers who care about animal welfare. In order to meet our side of the "ancient contract" we must rethink the many systems that involve regular painful procedures, such as castration and tail docking or involve putting animals for long periods in situations where they are bored or continually bullied by their peers without a route of escape.

At FAI we recognize that making fundamental change requires considerable investment, may take some time before the system is perfected, and even longer before it can be considered demonstrably robust. "Robust" when applied to an agricultural system means that it must operate without serious problems through all weather types, staff changes, holiday periods, and any other potential periods of disruption until the developers of the system have complete confidence and can convey this to others. In our experience this takes at least 2 years. Over the past 5 years we have developed a system for keeping pigs that addresses the two major animal welfare issues of pig production, namely the current need to dock tails and the use of the farrowing crate for sows.

Tail docking in pigs

As far as the pig is concerned, the major problem is not actually having its tail docked or cut off, although this is certainly a painful process (we know from human amputees that a stump can be hypersensitive). The major problem is what the need for tail docking tells us about lives of the pigs. Tails are cut off because, if they are not, other pigs start chewing them. Studies of pigs kept in environments in which there are plenty of things to chew spend about 60% of their waking time using their snout and mouth for investigation (Stolba and Wood-Gush, 1984, 1989). When pigs are housed in barren environments, where there is little of interest to investigate, the most interesting item becomes another pig. Thus the investigative behavior becomes "redirected" to other pigs. We see activities such as tail biting, ear chewing, and general rooting against the belly of another pig which are not seen in the wild. It is not difficult to imagine how interesting a tail would be (particularly a long curly one) in an environment that is otherwise devoid of much else. By enriching the environment of pigs by providing items which they find interesting, such as straw, woodchip, and other substrates, we can reduce the likelihood that they will chew other pigs' tails and so reduce the need for removing those tails in the first place. This is why understanding of which substrates pigs choose to investigate and chew in their natural environment is so important.

"Toys for pigs" have received considerable media coverage (Scott et al., 2006) but to provide just anything for pigs to play with misses the point that pigs are complex animals looking for some kind of reward from what they are investigating. "Toys" will not hold their attention for very long unless a reward is forthcoming. Rewards in pig terms are things such as food (the size of the morsel does not seem to be important and a small piece of barley grain found in straw is deemed a great prize), something to manipulate such as a piece of twig, and material to carry about and possibly mould into a nest. "Toys," on the other hand, are items such as chains and tyres which do not provide any reward and are thus rapidly discarded and ignored. At FAI we have found that woodchip (shredded waste wood) maintains the interest of pigs for literally weeks on end. They root around in it and never seem to tire of it. Woodchip also has the great advantage that it is an unwanted waste product that can often be obtained cheaply or even free from local councils and tree surgeons.

Using these ideas of what the pigs themselves want from their environment, we developed a very simple system which provided for all their basic needs. Although rooting is a key activity pigs also need to have "friends," other pigs which they know and are therefore less likely to fight with, they need to keep warm (straw provides this) and to get cool (a woodchip wallow under the drinkers provides this). Pigs can only lose heat by the latent heat of vaporization of fluids from the skin surface and being able to get wet is of vital importance to the pig when temperatures go above about 15°C. Our pigs are kept together in family groups, without any mixing, up to about 7 weeks of age and thus develop their hierarchy when they are small and can do little harm to each other. After weaning they are then kept in groups of up to 50 and have enough space (they are given three times that required by law) to indulge their various activities such as lying, defecating, feeding, and playing in separate areas which means they are not constantly disturbing each other. Using this system we have not had to dock the tails of our pigs for 5 years now.

When described in a few paragraphs like this, system development sounds very simple, but we know from a comparison with other current systems that each element is important, and if it is not there the system does not work. It is not always easy to provide good substrates (straw and woodchip) as these can clog up vital slurry systems. It is important that unfamiliar older pigs are not mixed together as they will invariably fight and it is important to keep stocking densities and group numbers low as these are key to making the best use of space and operating disease prevention programs. Many current systems do provide enrichment in the form of something added on a daily basis such as a rack of straw or a box of woodchip in one corner. The danger with such small scale additions is that a busy farmer might leave this job out when there are more urgent pressures and not filling up a straw rack or woodchip bin for just one day could lead to an outbreak of tail biting. The most effective method is, therefore, to make the enrichment an integral part of the system and then the whole system becomes more robust.

Persuading farmers to adopt the new system may require them to make a complete change in their philosophy. Many are still convinced that tail docking is essential. A further block to progress is that a farmer may not know which type of system the piglets are going into. Pig's tails are docked, if at all, when they are a few days old so unless a farmer is sure the piglets will end up in a system where tail-biting is unlikely to occur, the best strategy, for the pigs' sake, is to dock all their tails. On the other hand, tail docking is a time consuming and unpleasant activity for man and piglet and there is now increasing demand for pigs with their tails on. A pig with a healthy, intact tail is living proof that it has had a good quality of life.

We have had no tail biting but the cost of producing such meat is 30% higher than the national average. With further developments it might be possible to reduce this differential but the fact remains that to provide pork with good welfare means that the base price of pork will inevitably have to rise.

A free farrowing system for sows

The farrowing crate is one of the last remaining examples of close confinement of farm animals. It is exactly what it says it is, a metal crate which allows the sow to stand up and lie down but not to move around. It was developed to confine the sow prior to farrowing (giving birth) to stop her lying on her piglets and crushing them. This it does very successfully. However, it also severely restricts the ability of the sow to express her natural nest building behavior and to bond with her piglets. Like the battery cage for laying hens, this system was developed at a time when we knew much less about the needs of animals. It is important to reiterate this point as the system solved a problem for farmers at the time and those involved were not aware that they were creating such a great problem for the sows. However the scientific evidence for the welfare problems caused to the sow in a farrowing crate is now very clear (e.g. Jarvis et al., 2004).

The other common system in the UK is to keep sows outside and for them to farrow in individual arks. These are the often half circle shaped objects, which can be seen dotted all over farmland in the free draining areas of the country, and these do allow natural behavior at farrowing time. However there is not enough suitable and available land with the right climatic conditions in the world to move all breeding pigs outdoors. Unsuitable ground, which is not free draining, increases the risk of pollution and makes it difficult to keep the piglets warm and dry. We therefore felt it was imperative to develop an alternative system.

The development of this system has been very much more protracted and difficult than that of the "keeping tails on pigs" system and I will describe some of the key steps and the problems we encountered to give a feel for how much investment and commitment is needed to make fundamental changes. First of all, we scoured the literature to find out what was known about the activities of a sow in the wild, what scientists had been working on, and what innovative farmers were up to. The seminal work of Alex Stolba and David Wood-Gush on the Edinburgh "Pig Park" (1984) showed that, given the opportunity, domestic sows would build a huge nest away from the main group and then keep her piglets there for about 10 days before reintroducing them back to the family group. Dale Arey (1992) "asked" sows "how much you want to nest build?" by seeing whether they would work to obtain nest material by pushing a lever many times to get it. Sows "replied" that in the period leading up to farrowing they would work just as hard for nesting material as they would for food. This result made the important point in that what animals want can vary with time. Nest material is important for sows but only in the brief time before they give birth. To see what other farmers had already done, I looked at some innovative systems in mainland Europe and visited some extremely well run outdoor units courtesy of Fred Duncan of Grampian Foods in Aberdeenshire.

What we learned from all this information and from our own preliminary trials is that sows have two major needs at farrowing time. First, they must have privacy when they farrow and for about 10 days afterwards. This separate time is vital to allow the piglets to become strong without interference from other members of the family group. Second, the farrowing environment is vital in that we must provide a nest which the sows will use, for the sake of the piglets who cannot control their own body heat until they are about 2 weeks old and thus require an external heat source. The best results were obtained by placing the ark on a yard and providing nesting material (straw) inside the ark only. In this way, the sow can mimic the environment of a wild sow's nest which is typically a huge mound of grass, branches, and any other vegetation she can find in the vicinity. Both the nest in the wild and the farrowing ark allow the creation of a warm environment, which keeps a fairly constant temperature for the piglets due to the presence of the sow, as she spends a lot of time in the nest when her piglets are young.

One of the mistakes we and many others before us made was to confuse free farrowing (free to move around and build a nest) with keeping sows in groups. If a group of sows do not have sufficient privacy when they give birth, they will visit each other at crucial times, disturbing each other and increasing the risk of the piglets being squashed. We were able to solve this problem by making sure that each sow had her own private ark and outdoor run area to give birth in. Another mistake we made was to build these arks originally as open topped boxes which,

ANIMAL NEEDS AND COMMERCIAL NEEDS

although they were in a barn, did not keep the heat in and allowed the piglets to get too cold when ambient temperatures were low. Also the barn environment defeated the object of the exercise as some sows would build their nest outside the boxes using the straw provided in the box. We found the best solution was to have arks as used in outdoor systems and simply place them outside on a yard. A further mistake was to give the sow and her piglets too much room outside the ark, where inquisitive piglets sometimes could not find their way back to the nest and again were chilled. We solved this by making the space outside the ark smaller with no nooks and crannies.

Because of these various problems, we went through times of being very despondent. Our whole aim was to improve the welfare of both sow and piglets and at times in the development of the system we knew it was actually worse than the average system. At a particularly difficult time, guidance came for me in the words of John Gardner who was then Chairman of Tesco who simply said "Ruth remember failure *is* part of progress." We did not give up and our perseverance has paid off.

We were also greatly helped by the research that showed that there may be a genetic component to maternal behavior and not squashing piglets (Jarvis et al., 2005). This is hardly surprising as in nature those females who do not squash piglets will have more offspring surviving to breed themselves. We therefore decided to have a breeding program in which we only bred from sows that had a good record of not squashing their piglets. As with the sheep, we could do this quite easily as we have a pure bred herd. Despite gloomy predictions from others that we would have problems with disease, our free farrowing unit has maintained what is known as high health status for 3 years. It is, however, a small unit of only 30 sows whereas commercial units typically contain over 200 sows. I have recently been very pleased to visit a unit at Bishop Burton Agriculture which is running a commercial sized herd under such a yarded free farrowing system and is obtaining excellent results. The exciting prospect is that if we can select sows from free farrowing breeding stock to place into this type of unit there is potential to exceed the commercial results of the farrowing crate and then both pigs and man will benefit. It is too early to give commercial figures from this system but predictions are that although costs will be higher the differential will not be prohibitive.

Where do we go from here?

Often in my work I am reminded of Ghandi's well known quote, "The greatness of a nation and its moral progress can be judged by the way its animals are treated." It is a source of hope to me therefore that the leaflet produced by our government last year, in response to the bicentenary of the abolition of the slave trade, made the following statement: "2007 is a chance to make a collective commitment that in another two centuries time no-one should feel the need to express regret for our activities today." At present, our treatment of many of the animals throughout the world who provide our food falls into just such a category, which with hindsight we certainly will come to regret.

The good news is that change is happening all over the world. There are both new and traditional production systems that do take account of animal needs. Farmers in developing nations may not understand animal welfare as a subject but their agricultural evolution has been different and we often find that they have not yet broken the "ancient contract." The sad fact is, however, that many of these farmers aspire to the cages and crates of more "developed" countries as they see this as the way forward. We need to dissuade them.

If we want to see real changes made, then the most powerful tool is the one that we, everyone of us, as consumers have each and every time we buy food. The responsibility for continuing with unacceptable systems cannot be laid solely at the door of the farmer, although obviously changes will have to be made to practical farming. Our job as farmers is to rise to this challenge and develop effective, commercially viable alternative systems that offer the consumer real choice. At FAI we aim to be the voice of the many farmers throughout the world who have contributions to make and who want to make a difference. Change on a large scale will only happen when farmers have the confidence that a new system will enable them to make a living. We need to know that it is robust under all circumstances and that people want to buy what we produce. The evidence is that this process, eloquently described by Colin Tudge as the Agricultural Renaissance (Chapter 16), has already started.

ANIMAL NEEDS AND COMMERCIAL NEEDS

9 Hard work and sustained effort required to improve livestock handling and change industry practices

Temple Grandin *is Professor of Animal Science at Colorado State University where she teaches courses on livestock behavior and facility design and consults with the livestock industry on facility design, livestock handling, and animal welfare. She is the author of* Thinking in Pictures, Livestock Handling and Transport, *and* Genetics and the Behavior of Domestic Animals. *Her book* Animals in Translation *was a New York Times best seller.*

When I was a student at Arizona State University, I read a quote on the graffiti wall on the art building that said, "obstacles are those terrible things that you see when you take your eyes off the goal." It was the early 1970s and I was just starting my work in cattle behavior and the design of handling facilities. I never forgot that quote. Too often I have seen people get so bogged down with fighting the obstacles that they lose sight of the goal. One has to develop ways to go around obstacles instead of just bashing against them.

Today, over 30 years later, I have achieved many goals to improve animal handling and slaughter. When I first started I would have thought that many of the goals would have been impossible. Half the cattle in North America are now handled in restrainer systems I designed for slaughter plants (Grandin, 2003). Curved race systems that I have designed are used around the world (Grandin 1980, 1990, 1997a, 2000a). My most recent project was implementing the animal welfare programs for auditing the suppliers of McDonalds, Wendy's International, Burger King, and Whole Foods (Grandin, 1997a, 1998b, 2005, 2006). This program has resulted in tremendous improvements in animal treatment at slaughter plants. Now at the age of 60, I spend a great deal of time lecturing and mentoring to students to pass my knowledge on.

Be a pioneer

When one starts out as a pioneer, people will often think you are crazy. In the early days of my career, I handled cattle in every feedlot in Arizona to learn as much as I could. I was one of the first people to ask why cattle moved easily through some race systems for vaccinations and in others they balked and backed up. When I got down in the races to see what the cattle were seeing, people thought it was ridiculous, but I discovered that shadows, reflections, and little chains that moved scared the cattle (Grandin, 1980). If these distractions were removed, cattle moved more easily. At the same time, Ron Kilgour in New Zealand had independently made some similar observations (Kilgour, 1971). Finding his paper in the library assured me that I was not crazy so I kept on working.

Both Ron Kilgour and myself were looked upon as eccentric and crazy by the mainstream cattle and sheep industry folks. Pioneers are often criticized and they have to be strong to stand up to that criticism. In the biography of Jane Goodall, Peterson (2006) states that she was severely criticized by established scientists when she presented one of her first important papers at a scientific conference. Fortunately she kept on going and never took her eyes off the goal.

There are a series of steps that occur when a pioneer goes from being viewed as totally radical to mainstream. Organic farming and animal welfare are two areas where ideas that were considered radical 20 years ago are mainstream today. Dan Murphy (2006), a long-time writer in the meat industry trade press, wrote about me: "Once dismissed as an eccentric (at best) then later accepted as an eccentric with some intriguing ideas on handling and stunning livestock and much later accorded all kinds of honors as a revolutionary thinker!" I became respected because the equipment that I designed, worked. People hired me because I was able to solve problems with handling and stunning of livestock that they could not solve.

Hard work

To bring about changes requires sustained hard work often for a period of years. Some people started to take my ideas on cattle handling seriously after they read articles I had written in cattle magazines. A few readers tried some of my recommendations for facility design and cattle behavior. The support I received from this small number of people who had successfully used some of my cattle handling methods motivated me to keep persevering. I discovered that many people were more willing to adopt the new design for equipment instead of a behavioral method for low stress management of cattle. In addition, it takes hard work to become really knowledgeable in both the science and the field work. I read hundreds of journal articles, popular articles, and spent weeks on ranches and meat plants to become a real expert.

During the last 30 years, I have written articles on basic cattle handling principles over 75 times. They have appeared in livestock publications around the world. I have also done several hundred speaking engagements and classes on handling cattle and facility design. Often I get asked by other scientists if I find this boring. Sometimes it is boring, but being a constant presence is required to change how things are done. Bringing about change requires sustained effort. Early in my career I accepted every speaking engagement and every invitation to write an article.

In my work on equipment design I spent many long days working with construction crews to make sure my designs were built and installed correctly. This supervision was essential to prevent installation mistakes that would cause a system to fail.

Early adopters must succeed

When a new technology is first introduced, one must make sure that early adopters succeed (Grandin, 2003). If the early adopters fail, the introduction of that technology could be slowed down for years. In the US this happened to electronic sow feeders for group housing. In the 1980s they were sold to hundreds of farmers before the gate design was perfected. The dissatisfied farmers threw out the whole concept and were unwilling to try new models until years later.

The steps for successful transfer of new technology or farming methods are:

1   Choose early adopters that believe in it and want to make it work. Choosing the wrong early adopters can cause a good project to fail.

2   You must spend a tremendous amount of time with your early adoptors to train them and avoid mistakes that could cause failure.

3   Perfect the method or technology before putting it out on the general market.

4   Communicate results in many different types of media for many years.

I spent days and months on the job sites of my early adopters of my curved race designs and restrainer systems. I lived with the construction crews and stayed in junky motels with them. On a typical day at one of the meat plants I got up at 5:30am so I could be at the plant at 6:00am. When my restrainer systems were first introduced, it took several months to work out all the problems in the new equipment. It was my job to figure out how to fix problems. Lunch was from a disgusting catering truck and I had to put up with screaming managers who thought that a totally new invention should work perfectly in the first 5 minutes. This work was highly stressful but also very rewarding. When the shift ended at 15:00, I had to stay and help the construction crew fix things. The construction guys were wonderful to work with. Together we brainstormed on how to fix the system. Many of them were very creative. If I was lucky I'd get back to the hotel by 20:00.

My loyal, early adoptor stood by me and defended me when the other employees in their plants were swearing and angry when something went wrong with the equipment. I always had to be nice. To keep my sanity when other employees screamed, I had to view them as if they were two-year-olds having a tantrum. To calm them down, I took them to the conference room and explained how we were going to fix the problem. The "tantrum" phase of new equipment start-up was not fun. Implementation of new technology on a farm is psychologically easier than at a large meat plant because there are fewer people involved.

Reward of hard work

The reward was seeing the final system working smoothly. Instead of shrieking, bellowing frightened cattle, the animals moved quietly through the system. Some of the most spectacular improvements were in plants that had previously shackled and hoisted live cattle prior to kosher slaughter. In the US, religious slaughter is exempt from the Federal Humane Slaughter Regulation and shackling and hoisting is still legal. In the 1980s and early 1990s, employee safety motivated many plants to eliminate shackling and hoisting of live cattle. Today large meat buying companies will not buy from plants that shackle and hoist live animals. Unfortunately there are some small US plants that still shackle and hoist cattle, calves, or sheep. In South America, shackling and hoisting, shackling and dragging, and other highly stressful methods of restraint for kosher slaughter are still being used. I remember one dreadful plant where 100% of the cattle were bellowing and thrashing as they were dragged and hoisted up by a chain around one back leg. I was hired to tear out this atrocity and replace it with an upright restraining box to hold the animal in a standing position during kosher slaughter. When I first visited the plant, the manager was clueless that hanging live 500 kg cattle up by one leg was cruel. He asked me what I thought of his system and I told him it was terrible and needed to be replaced; which he agreed to do. One of the things that was hard about these projects is I had to act nice even though I was really upset about how the cattle were treated. I had to keep my mouth shut so I could achieve the goal of changing the equipment.

A few months later I returned with the construction crew to install the new upright restraint box. It took the equipment company several months to order materials and build the new equipment. The kosher box was specially designed to prevent excessive pressure from being applied to the animal (Grandin, 1992). Many restraint boxes were poorly designed and they applied excessive pressure which caused struggling and vocalization. It also had a nonslip floor because animals panic if they feel like they will fall. Even slight slipping makes an animal become agitated.

A good equipment start-up

During the start-up, I ran the hydraulic controls. Each steer entered quietly, and there was no bellowing. He walked in calmly and I nudged him forward with the pusher gate and held his head with the head holder. I had worked hard to get the lighting in the box just right. Animals tend to move from a darker place to a brighter place (Van Putten and Elshof, 1978; Grandin, 1982; Tanida et al., 1996). There was a light over the head holder to provide indirect lighting to attract the animals in. I had also installed a solid barrier around the head holder so that the animal did not see people and other distractions. Controlling what cattle see is essential to prevent balking and backing up (Grandin, 1992, 1996, 2000a). This avoids the balking problems that Ewbank et al. (1992) described when head holders are used.

It was amazing how well it worked. I forgot about working the controls and it was like reaching out with my hands to hold the animal's head. The instant he was held, the rabbi made the cut before the steer had a chance to fight the restraint. Now that the steer was held in a gentle manner, I could determine his reaction to shechita. I was surprised that he did not appear to feel the cut. Even when I put the head holder on so lightly that the animal could pull his head out, he did not move it. The special long kosher knife is essential. I have observed slaughter without stunning of cattle done with short knives and the animal violently struggled due to pain. To maintain an acceptable level of animal welfare during kosher slaughter requires very careful attention to restrainer design and operation. In eight plants with upright restraint equipment, I worked with them to correct problems with their kosher restrainer, such as excessive pressure, slipping on the floor, or sharp edges. To reduce stress on the animal, the cut was made within 10 seconds. In seven out of the eight plants where I worked on their upright restrainer, 5% or less of the cattle vocalized in the leadup race or in the restrainer. Before the modifications were made on the upright restrainer boxes, the worst plant had 32% of the cattle vocalizing due to excessive pressure exerted by the rear pusher gate. Vocalization is a sign of stress in cattle (Dunn, 1990; Grandin, 1998a).

Applying knowledge in new ways

From this work on design of cattle restrainer devices, I developed four principles of animal restraint. They work for holding many different species ranging from cats and dogs to cattle and horses. I teach these methods at our veterinary school and it was really rewarding when one of the students told me that when he interned at a veterinary clinic they let him handle all of the difficult cats. He was known as "the kitty whisperer." He had applied my techniques. These methods work both for large animals held in a restraint device and small animals held by a person's hand.

Principles of restraint

1   *Avoid triggering the fear of falling* – If the animal is standing, it must have a nonslip surface to stand on. Even slight slipping triggers panic. If the feet are off the ground, the animal's body must be fully supported.

2   *Smooth, steady movement of a person's hand or an apparatus is calming* – Sudden jerky motion of a person or an apparatus frightens animals.

3   *Use optimal pressure* – Just the right amount of pressure. Firm pressure over a large area of the body is calming. It must not be either too tight or too loose. The most common mistake is squeezing animals tighter when they struggle.

4   *Block vision (grazing animals only)* – Blocking the vision either with a blindfold or solid sides on a race keeps wild animals calmer. This is especially important for animals that are not completely tame.

Give away information

Throughout my career, I have given away tremendous amounts of information on cattle behavior and facility design. It was specific, practical information that people could take home and use on their operations. Many people tell me not to give it away. There are some other people working on animal handling but they have had less of an effect because they want to sell all their information. I make a living on consulting fees, my part-time professorship at Colorado State University, and speaking fees. Making a living by selling books and videos on cattle handling information is impossible. It is just too specialized. By giving information away, I get more consulting jobs than I can handle. Another mistake that people make is getting over capitalized with too many employees and expenses. I have seen a lot of people go broke because they are set up with a big office and far too many employees. For most of my career, I have worked alone. Today I have only one employee and two people who do part-time work on my webpage and manuscript preparation. Also today I give information away on my webpage (www.grandin. com). I do not make people register or pay fees because I want maximum distribution. People want the information because it works. Before the internet was invented I gave away drawings and lots of information in magazine articles.

When your information is freely available, other people will use it when they need to develop welfare or production standards. The information that you give away may become the new standard for your country or another country. Another thing I did with much of my design work was to place it in the public domain so nobody could patent it. Many good technologies are not being used because they are tied up in patent disputes. When I invented new things, I published the drawings in a magazine that had a verifiable date which destroyed world patents. At the time I put my restrainer designs in the public domain, I was making a good living on consulting fees and fixing all the systems that people installed in the wrong way because they did not follow my instructions.

Be positive and never nasty

Some advocates are not very effective because they attack everybody and everything. Instead of attacking, one should present positive better ways of doing things. The people who are nasty do not realize that they have taken their eyes off the goal. When I first started writing in farm magazines I was disappointed that I often received no mail about many of my articles. However, when I saw a reader at a cattle meeting he would tell me how much he liked the article or that a facility design I wrote about worked on his ranch. Unfortunately, when somebody hated something I wrote, that news got back much more quickly than the opinions of many positive people who were less inclined to take the time to respond. When I got the first nasty letters to the editor in the early 1970s I was upset. Today in the age of e-mail it is even easier for angry people to respond. I had to learn never to respond in an angry manner. Recently I responded in a very diplomatic manner to a mean review of my book in an animal rights publication. By being diplomatic, my response was published and probably had a good effect on some readers in the animal rights community. Another thing I had to learn is that everything I write and everything I say in a public forum or to the press is for publication to the world. Many people are losing jobs and getting in trouble for nasty or inappropriate e-mails. If you are upset you should pick up the telephone and talk to a trusted friend. Do *not* use e-mail or write angry stuff on blogs or chat rooms. Electronic communications are often archived and they can come back to haunt you.

Communicate with a wide audience

To make a change one should communicate to a broad audience on many different levels. I have written up my work in *both* the scientific journals and the farm press. Information also should be made available on the web that is easily accessible to people who do not have subscriptions to journals or access to a university computer. Recently, Catherine Johnson, my co-author of *Animals in Translation*, found it impossible to find papers from certain scientists that she could access without having to pay a fee or have a subscription. It is essential for scientists to post a few basic papers on fully accessible free sites.

Throughout my career I have made an effort to communicate my knowledge on livestock behavior, facility design, and humane slaughter to many different audiences. I communicate to producers, scientists, animal NGOs, engineering companies, and students. There is a tendency in our profession to talk between ourselves or to avoid talking to people that one may consider the enemy.

Many people in the livestock industry refuse to talk to people in an NGO animal advocacy organization. I talk to them all the time and I call it "keeping the embassy open for dialogue." Within any large organization there will be individuals within it where there can be dialogue. I may be on the opposite side of the fence but there are areas where we can cooperate and areas where we can agree to disagree. Communication helps to prevent destructive polarization. On many issues there is a middle ground of opinion in-between the radical views of either the extreme right or left. An article in *Nature* stated that there is a need for people with moderate opinions on the use of animals in research to speak out (Marris, 2006). Doing this is difficult because you risk getting attacked by radicals on both sides of the issue. One of the things that has made me successful in bringing about change is that today I have become an outspoken moderate who communicates with a broad range of people. Getting attacked is very unpleasant, but I have to remember that 90% of the population disagrees with the most radical views on either side of an issue.

Open the door and show

I am proud of the systems I have designed and I have taken many people who have never been to a slaughter plant to see my systems in operation. Many people are amazed at how calmly the cattle walk up the race. For years I have preached to the livestock industry that we need to open up the door and show the public what we do. A well-run slaughter plant will pass the test. However, extreme confinement such as sow gestation stalls and veal crates will not pass the test.

Confidentiality to maintain access

In order to gain access to so many different places, I keep the names and locations of many places secret. When I talk about the industry or write about it and discuss bad things, I keep the source confidential so I will still be able to have access. Over the years I have described all kinds of bad and abusive things in detail but I do not say where they happened. I tell anybody who wants to know exactly how farms or slaughter plants work, but I do not say where they are. I will tell you everything except "where."

One must remember never take your eyes off the goal. If I revealed the location I might win a battle but I would lose the war and be kicked out of the industry.

I have another policy that has served me well. Highly confidential discussions with livestock producers, restaurant chains, supermarkets, meat plants, and animal advocacy organizations are done only on the telephone or in person. Never use e-mail for sensitive negotiations. Everything I write including a consulting report for a client is written as if it was going to be shown to the world on the internet. We are becoming more and more of a world of lawyers and there are certain private negotiations where no written records should exist. It is during these negotiations that great work often gets done to bring about huge positive change. There are other situations where a paper or e-mail trail is essential. I keep all drawings and letters on design so I can prove that I included important safety features.

Avoid off-topic controversies

My goal is to improve the handling and treatment of livestock during handling, transport, and slaughter. I am a great fan of the original classic *Star Trek* and they had a prime directive, to not interfere with life forms on other planets. For the last 35 years I have been careful to not get off task and let other controversies interfere with my "prime directive" of improving animal handling. I call this being "project loyal." For example, if I am at a farm building a new system I must avoid getting into a fight with the owner over politics. To avoid problems, I keep my views on national politics and other sensitive topics to myself. Recently at a large lecture I was asked about my opinion of the president. I replied that this was not the proper venue for me to answer that question and to talk to me out in the hall in private.

Importance of objectivity

For years I have had a reputation for being very objective and separating fact from opinion. Recently I was one of the teachers at an industry sponsored course on animal welfare at meat plants. In the evaluations one of the students commented that I was the only instructor who gave totally objective information on different stunning and handling procedures. When I communicate with the public, I emphasize the importance of finding out what is happening in the ground out on a farm instead of just reading ideological literature. If you are going to get involved in any issue you must read literature from both sides. Avoid the rhetoric and read scientific literature and articles written by people who have "hands on" knowledge of the issue. I read a wide variety of scientific journals, livestock magazines, and animal advocacy literature. You must know what is happening out in actual farms and plants.

Technology does not solve all problems

When I first started my career I thought I could fix everything if I could just design the right system. To my dismay, I found that only about 25% of my design clients operated and maintained equipment correctly and there was a bad 25% that wrecked the equipment and abused animals.

One of the greatest turning points in my professional life was implementing the McDonald's animal welfare auditing program during 1999. I was hired by McDonalds Corporation and Wendy's International to train their food safety auditors to audit the slaughter plants for animal welfare. These audits forced plant management to start managing the animal handling and stunning in their plant. The process of implementing the program was amazing. The food safety auditors took a great interest in it and became diligent about doing the audits. There is more information in Grandin (2000b, 2005, 2006). The tremendous purchasing power of these companies changed the meat industry within a few years. But a word of warning is needed. Constant vigilance is required to maintain standards. The plants must be continually inspected.

Legislation is not always the answer

Legislation is only one way to bring about improvements. The other is the tremendous purchasing power of major buyers. This purchasing power can bring about change. To make change occur, two types of people are needed – the activists and the implementers. I like to use a metal working analogy. Activists heat and soften the steel so that implementers like me can bend the soft steel into new shapes. This is exactly how the McDonalds audits started. The work of activists softened the steel and I implemented the program. There is a shortage of good implementers who have good practical field experience for programs. To be successful, programs have to have practical methods that will work. Without good practical field people, a well intentioned program may be a failure. There is a huge need to get young students interested in field work. In the US, many students who are interested in animal issues are going into law. Lawyers are good for "heating the steel" but they are poor implementers and will not be able to "shape the steel."

To bring about the greatest amount of reform and constructive change, the most effective activists know when to back off after some good change has occurred. In one case some big improvements occurred in a slaughter plant when a new manager was hired. It was very discouraging for the new manager to continue to see his plant bashed by activitists on the internet. Continuous criticism on the internet after changes have been made may cause a good manager to become discouraged.

Contact with the field

It was extremely gratifying to take the executives from major restaurant chains and supermarkets on their first field trips to farms and slaughter plants. Before these trips, the animal welfare issue was just an abstract concept that the executives gave to the legal or public relations department.

When they saw a well managed slaughter plant, they remarked that it was much better than they thought it would be. However, when they saw something really bad, such as an emaciated half-dead dairy cow going into their product, they were horrified. This motivated them to implement their programs. In 15 minutes I observed executives who switched from viewing welfare as an abstraction to a reality where action was needed.

Achievable practical goals

To make effective changes take place, the goals must be practical. Sometimes it is better to get 80% of what I want to achieve and actually get it, than to go for 100% and not get it. Sometimes a door opens and an opportunity arises to really bring about change. A collaboration between enlightened progressive people in both the US Department of Agriculture and the American Meat Institute made it possible to create the 1997 Good Management Practices Guide that became the basis for the very effective restaurant audits. Dr Bonnie Buntain at the United States Department of Agriculture (USDA) and Janet Riley at the American Meat Institute (AMI) were highly supportive. Dr Buntain was instrumental in funding a USDA survey where I collected data to form the basis of the objective welfare scoring system (Grandin, 1997b, 1998b). Janet Riley and her progressive members at the AMI adopted the audit guide that I had written (Grandin, 1997c). When the restaurant audits started, the AMI guidelines provided a clear objective method for evaluating the treatment of animals in the plants. It was a strict standard that had credibility. Unfortunately many producer groups create standards that allow the farms with the worst practices to pass.

Eliminate vague guidelines

I constantly fight vagueness. When writing guidelines, the words "sufficient," "proper," and adequate should be banned. What one person would call proper handling, another person could call torture. Another feature of the guidelines is that they followed the HACCP (Hazard Analysis Critical Control Points) model used for food safety. The principle is to pick out relatively few critical control points or core criteria that must be passed to pass the audit. This avoids the bad situation where a plant may get enough points on record keeping and training documents and poor stunning would be allowed to pass. When objective numerical scoring is used, a plant has to attain certain numbers to pass. The system is fully described in Grandin (1997c, 2005, 2006).

Each animal is scored on a yes/no basis. To pass the audit, a plant must receive a passing score on all five critical control points. The five critical control points that are measured are:

1 Percentage of animals rendered insensible with one application of the stunner.
2 Percentage rendered completely insensible before hoisting. Must be 100% to pass.
3 Percentage of animals that vocalize (moo, bellow, or squeal) in the race and stun box. Each animal is scored as either vocalizer or silent. Lairage vocalizations are not scored. For more information on vocalization scoring, see Grandin (1998b, 2002). Vocalization is associated with stress in cattle and pigs (Dunn 1990; Warriss et al., 1994; White et al., 1995; Weary et al., 1998).
4 Percentage that slip and percentage that fall down during unloading, driving, and restraint. Score slips and falls separately.
5 Percentage moved with an electric prod.

These five critical control points measure outcomes instead of telling people how to do things. There is also a list of abuses which would result in an automatic audit failure. They are dragging animals, beating, poking sensitive parts such as the rectum, and deliberate rough handling such as slamming gates on animals or driving animals on top of each other.

My USDA survey was used to help determine baseline scores before the audits started. The limits were placed so that 25% of the plants could pass and the others would have to work to achieve the standard. Many slaughter plants failed their first audit and then they were given time to bring their operations up to the standards. A common mistake is to set the limits too low so that even the worst producers or plant could pass.

Plant management and employees became very inventive to come up with better tools for driving livestock to greatly reduce electric prod usage. To improve captive bolt stunning, the main thing the plants had to do was improve maintenance of the stunner. Fortunately in many slaughter plants, much higher welfare standards could be attained with simple improvements such as:

1   For captive bolt, improved stunner maintenance was essential. For electric stunning, tong placement and bleeding had to be improved.

2   Training employees in behavioral principles of livestock handling.

3   Nonslip flooring was installed in high traffic areas such as unloading ramps, races, and stun boxes.

4   Installation of solid sides to block the animal's vision in races and stun boxes.

5   Lamps were added to light a dark race entrance or lamps were moved to eliminate reflections on wet floors or shiny metal.

On farms, the same HACCP approach should be used. For example, on a dairy if there are too many lame or skinny cows, the dairy should fail the entire audit. For all species, if animals have wet muck for bedding, the audit should be failed.

You manage what you measure

The most effective audit systems measure outcomes or performance with objective measures which can be directly observed by an auditor. Audits that mostly rely on paperwork are less effective. Many different bad practices cause lameness. Instead of measuring these inputs, the outcome that is lameness is measured. Some people have criticized the auditing program because it allows producers or plants to not be perfect. For example, an audit done by a restaurant could be passed if 1% of the animals fall during handling. In plants where falling is audited on a regular basis, the percentage of animals falling down may drop to one in several thousand. Another example is vocalization scoring. Before the audits started, the worst plant had 32% of the cattle vocalizing. Today the worst vocalization score in the audited plants is only 6% and the average score is 2% (Grandin, 2005). For stunning, the baseline USDA data indicated that only 30% of the plants were able to stun 95% of the cattle with a single shot. Today over 90% of the plants are able to do this (Grandin, 2005). The best plants also have a good system of conducting weekly internal audits, and a plant manager who cares about animal welfare. Objective numerical scoring enables a farm or slaughter plant to determine if their practices are getting better or becoming worse. It helps prevent "bad from becoming normal." Lameness in dairy cows is a good example. Over many years, it kept increasing. Since it was not being measured, people did not realize that it was slowly getting worse.

Suggestions for the future

Getting students interested in working in the field and solving practical problems is essential. I have a few good field work students, but unfortunately many people do not want to work as hard as I have. Working in the field was often stressful, uncomfortable, and tiring but it was also greatly rewarding.

To be successful in making change requires the following:

1   Knowledge of BOTH the practical and scientific aspects of your area of specialty. One must keep current by continually visiting farms and plants. Continuous reading of the scientific literature is also required.

2   Continuous writing and communications of ideas and methods. Similar lectures and articles must be done many times to many different audiences. Build a website full of practical information with free access.

3   Steady sustained effort over a long period of years.

4   The advice must be practical and it must work on the farm or in on the plant.

5   Be willing to be criticized especially when doing new pioneering things.

6   Have the wisdom to work on changing things that can be changed and do not bash your head against the things that cannot be changed.

7   Be positive in your communications and always keep your word. Present the SAME message to the livestock industry, the public, and animal advocacy groups.

BRINGING ABOUT CHANGE

10 Bringing about change: a retailer's view

Michelle Waterman *currently works as the Senior Commercial Manager for Agriculture at Tesco and has responsibility for all their livestock standards and farming communications. She has been with Tesco for 7 years, and previously worked as an Animal Health Inspector for Surrey Trading Standards. She took a degree in Agriculture with Land and Farm Management at Harper Adams, University College. She farms 400 acres in Surrey with her partner, where they have sheep, cattle and a few poultry and are in the process of converting to organic. She has always had an interest in animal welfare, particularly in how a large business can positively influence the industry.*

Working as Agriculture Manager for one of the largest food retailers in the world puts me in an almost unique position to drive change in animal welfare. My job is to source animal products for our stores which meet our customers expectation of animal welfare, at the same time as ensuring we have the most competitive sources of these products available. We have a "Livestock code of practice" for every species that we procure, detailing all aspects of production, feeding, transportation, and slaughter. We implement these through farm assurance schemes, our own internal audits, and the use of third party dedicated unannounced audits. We can use these codes to drive change and also to highlight through our "aspirational" standards where we believe the industry needs to focus going forward. As a business we have committed to our customers, and British farmers, that where we import products for sale in the UK, they will be expected to meet "equivalent" production standards. By equivalent I mean that the same outcome will be achieved but the methodology might be different. For example the type of poultry housing throughout the world varies in relation to the climate. This has meant that our standards are now applied in many countries around the world.

My job involves working with all parts of the livestock supply chain. In the field of animal welfare, I talk to scientists and researchers about their work and hear about their latest research findings on animal welfare and health, and I talk to farmers and our suppliers/processors about the problems they face in keeping and sourcing animals and selling their products. Generally speaking, research is often poorly communicated to the wider industry and our suppliers are resistant to change and quick to point out how difficult it can be to implement the research findings in practice. So my job is to bring these two groups together, trying to improve animal welfare in ways that are possible for farmers and still allow them to be profitable and produce livestock in the numbers that we require.

In practice, this is often hard to achieve, largely because we have underestimated the sheer difficulty of implementing change on the ground. We now need completely new ways of approaching the genuine problems that stand in the way of progress and in this chapter, I will try to outline what some of these are. Bringing farmers and suppliers on side, so that they see for themselves the advantages of improving animal welfare, is the key.

The first thing to say is that change is difficult. If people have always farmed in a particular way, they often see it as much easier to keep doing it that way, rather than to invest in new systems or risk making things worse than they already are. We have to remember that the sort of change we are talking about could mean radical changes to the whole way in which farmers rear their animals and run their farms. This means that even if I become convinced that something needs to be done to improve animal welfare, I can often meet considerable opposition, even from farmers who care deeply about their animals. They either don't believe it would improve welfare, or that it wouldn't work, or that, even if it did work it would cost too much and put them out of business. Sometimes they can even back up their arguments with their own bitter experience – they have already tried it and it didn't work.

In the face of such opposition I then have three options. The first option is to insist, through our livestock codes of practice, that a standard is essential. This is what we do where we believe that compliance would be an absolute expectation of our customers, for example, requirements that all animals must receive pre-slaughter stunning, an absolute limit of 8 hour transportation time prior to slaughter, and that all holding must be audited by an independent farm-assurance scheme. If these are not implemented, we will simply not be able to purchase products from a supplier. Achievement of such change is thus through the full economic influence that a large retailer has. This approach has its advantages. It gets things done.

But it can also lead to a half-hearted "tick-box" approach. A good example of where this has happened is in the field of veterinary health planning. From the outside the concept of the farmer agreeing a plan with his vet for the year ahead, to ensure preventative rather than reactive measures are implemented (e.g. vaccinations, improvements to building design, disease profiling, etc.), would seem sensible and one that would be readily taken up. Since the farm-assurance schemes adopted this as one of their standards, a health plan does now "exist" on every farm.

Unfortunately though, only a percentage have become the potent tools in improving health (and therefore welfare) on farm that they should be, either because they are only reviewed the day before the next annual audit, or the quality of input from both the farmer and vet in the first place was insufficient to bring about change.

Examples like this show where we need to embrace new techniques to bring about meaningful improvements in the way animals are produced. I find myself half way between insisting on what I think is right and backing down in the face of objections. I then have to try to get farmers on board by convincing them that the proposed improvements to animal welfare are not only possible and desirable but would actually be good for business too. If they buy into a system because they can see the advantages for themselves and their animals, they are much more likely to make it work than if they are compelled to do so by a retailer, or legal requirements they cannot see the point of. Persuasion – working with farmers to bring about change – is the most promising way forward and the most likely to achieve genuine change and to make sure that change is sustained. It is also the most difficult to achieve, which is why I want to go into it in some detail.

The example I shall use is a world-wide animal welfare issue: tail docking in pigs. Almost wherever pigs are kept, the tails of young piglets are cut off shortly after birth without anesthetic. The reason given for this mutilation is that, if it is not done, the groups of growing pigs will almost inevitably develop the extremely damaging habit of biting off each other's tails. Tail-biting and aggression are very common in groups of growing pigs and so over 90% of the pigs we currently buy are tail docked, often even those that come from systems deemed as higher-welfare, such as "outdoor-bred" or "Freedom Foods."

The actual act of tail docking is probably not that painful; farmers report that the piglets don't even stop suckling while it is carried out. However, the reasons we have to do it in the first place are the ones that need to be addressed. Rearing large numbers of pigs, in often fairly barren environments, where they cannot express their normal behavior – rooting in the ground – means they often end up fighting each other and outbreaks of tail-biting occur. This is indeed a massive welfare problem for the individuals involved, but the real issue lies behind this: keeping the pigs in these barren environments in the first place. If we can rear pigs without the need to dock their tails it will prove a good indicator (or outcome-measure) that we have addressed the much wider issue of providing these animals with an environment that meets their needs.

Faced with this immense opposition to change, often from people who have a lifetime's experience of raising pigs and speak from bitter experience and disillusionment, what am I to do? It is clear that I cannot go for option 1, which is compulsion. There are not enough farmers around willing to change their systems and risk not tail-docking their pigs. That leaves persuasion and the long haul to convince farmers to do something most of them are convinced is difficult to do, even for animal welfare.

The starting point is the knowledge that it is possible to keep pigs successfully – even under intensive conditions – with their tails left on. This knowledge comes from scientific research which has shown, in a variety of circumstances, that tail-biting and aggression do not occur or occur at such low levels that they are not damaging. One of the scientists who was most influential in showing this was the late David Wood-Gush to whom this book is dedicated, and there have been other scientists who have since confirmed his work. But, of course, small scale studies done in universities or schools of agriculture without the pressure of commercial farming, are never going to convince hard-pressed farmers that these findings have anything to do with them. They would put more weight on their own experience of seeing tail-biting happening in their own pigs than they would in any academic study. That was why the setting up of FAI (Food Animal Initiative) was such an important step. Although close to Oxford University, FAI are commercial farmers, paying full commercial rent on a tenant farm and in no way subsidized by the University. In fact, they pay money into the University, an arrangement that is envied by other UK universities struggling to keep loss-making university farms going.

FAI were thus set up as a "half-way house" between academic research and farmers. Being commercial farmers themselves, the idea was that they would be more likely to have influence over what other farmers do, than academics or retailers are. As described in more detail by Ruth Layton in her chapter, one of the first projects FAI undertook was to keep pigs with their tails on. And it worked! The FAI pigs had very low levels of aggression and no tail-biting at all, even though every pig tail was intact. What was the secret?

Fortunately, the system was very inexpensive to set up and the key ingredients were very simple. The piglets were reared communally in the same peer group all their lives and so did not experience the social mixing after weaning that most intensively housed pigs go through. They were weaned from their mothers relatively late and they had plenty of straw and woodchip to sleep, play and root in. They were not aggressive to other piglets because they knew them all from an early age and they had plenty of genuinely interesting objects to manipulate other than the tails of other pigs. The pigs were also tested and found to be free of the major diseases that have plagued the pig industry over recent times.

The fact that FAI were commercial farmers and had achieved success with relatively small changes (bedding, stable family groups) convinced me that the system could be much more widely used – but would need to be adapted to work in other systems. But I also began to realize that just telling farmers this was not going to be enough. The fact that they had tried and failed meant that the system had to be right and that there were some essential ingredients that they needed to be told about to make sure that it worked this time. Some tails-on systems work and some don't and it was essential to be able to identify the key differences. Also, although large numbers of farmers came to look at the FAI system, very few ever implemented it for themselves. Rollout was going to be a lot more complicated than just having a single demonstration unit at FAI. Every farm is different and the concepts would need to be adapted and modified to suit each farm's unique design.

Although FAI was originally set up as a "halfway house," it has become clear that, in practice, it was only step 1 on a rather longer path to persuading farmers to adopt new practices. Although financially distinct from either Oxford University or their core sponsors, the FAI farm was seen by "ordinary" farmers as not quite ordinary enough. It had media attention, academic attention, commercial attention, whereas they had none of these things. As a demonstration farm, therefore, it provided an interesting day out but not the final spur to go home and do things differently for themselves. Another step was needed so that FAI should really be seen as a "third of the way house"; an essential part of convincing me, as a retailer, what was possible commercially, but needing step 2 to complete the process of rollout.

What I mean by step 2 is providing the face-to-face, on-the-spot work that Temple Grandin has already described so graphically (Chapter 9). It means going to a farm and being on hand to deal with problems as they arise and before they become serious. It means pointing out to a farmer what he is getting wrong and where he is going right. It means being a mixture of a trouble-shooter and an agony aunt so that the farmer doesn't give up just because things don't go right at first. Then, when he can see what a large difference attention to small details can make, he knows himself what to look for and what to change and, with any luck, will have bought into the system himself. But, as is apparent from Temple's description of what she has had to do to get things up and running, this is an immensely labor-intensive process. So one of the ways in which I believe we, as a major retailer, can really make a difference is to supply that expertise in the form of trained consultants to go out and help more farmers through the difficult initial period to make sure they have not just got the idea of a new system, but understand all the practical details that are necessary to make it work.

Of course, we need farmers ready and willing to be pioneers or "champions" of a new system such as the FAI tails-on-pigs system, but my experience so far is that this is not as difficult as it might sound. Although the pioneer farmers do not receive any direct financial benefits from us (that would negate the whole purpose of their being "ordinary" farmers), they do receive help, advice, and trouble-shooting. Someone goes in and gives them an external view of their business and of their animal welfare. In return, we ask them to be willing to let other farmers come and see what they are doing and to give them information about the costs they incur and the benefits they derive form adopting certain practices. Obviously, we don't ask them for all the details of their business, but other farmers want to know whether key investments pay off, for example, whether investing a certain sum in a veterinary health plan results in savings in veterinarian's bills in the long run.

BRINGING ABOUT CHANGE: A RETAILER'S VIEW

As the retailer buying their produce, we are genuinely interested in how they get on, what works what causes problems, and what the extra costs of the system are. Regional producers clubs could assume much more importance than they have now and become a forum in which farmers exchange ideas about practical problems such as keeping animals healthy in the face of the ban on antibiotics, how to provide bathing water for ducks and still keep the straw bedding dry, and so on. I believe that it will take only a relatively small number of pioneer farmers to have a large effect on the rest of the farming community. Our current aim is to have about 20 farmers per species as "champions" and that once other farmers have seen what works, this will have a trickle-down effect on the rest of the farming community. If the others see that it doesn't cost a fortune to change and that there are genuine business advantages in achieving higher standards of animal welfare, then we will have a critical mass of farmers willing to implement change. At that point, I can move an improvement in animal welfare into a standard within our livestock codes of practice, which must be complied with wherever we source from. Any that are unable to comply after that will no longer be able to supply.

Although I am fully committed to this two-step approach, I certainly do not underestimate the problems that still exist or the importance of other factors in bringing about change. Two of the most important factors, both intimately linked, are cost and what customers want. Improvements in animal welfare often do cost money and someone, either the customers or the producers or the retailers, are going to have to pay for them. A further advantage of having both FAI and our champion farmers involved is that the cost implications are known about and fully documented.

This doesn't solve the problem of who pays for them but at least everyone is clear about whether there are net savings or net costs. Sometimes, as in the case of investing in good veterinary health planning, there should actually be savings. At other times, demands for say, extra space, will inevitably lead to a product being more costly.

In some cases, for our "premium" ranges, where people have enough disposable income to be able to pay for higher welfare standards, the product costs more and people pay more accordingly. For them, good animal welfare is enough of a priority that they will pay more and we can justify the higher costs of these products. In fact we are often responding to the demands of our customers. Everyone understands (or at least has an idea of) what "free-range" means when applied to eggs or meat chickens. It means birds that have access to the outdoors so that they have grass, and natural daylight and fresh air. We can show pictures and explain the welfare benefits relatively easily. For this reason, companies that specialize in top of the range products can sell themselves on being "all free-range" or "all organic" and have higher cost structures by appealing to their relatively welfare (or at least ethically) knowledgeable customers.

For retailers that attempt, as we do, to appeal to a much wider market and improve standards not just at the top end but the full range, the situation is more complicated and it may be much more difficult to sell products on their welfare benefits. Not everyone is prepared to pay more for high welfare and many of our consumers are more driven by other product attributes such as quality, price, packaging, simplicity of preparation, provenance (where the animals come from), etc. In other words, for a great many people, animal welfare is not a consideration when they shop for food. That said, our customers trust the Tesco brand, and as such expect that we will have ensured that the production system would meet their expectations. This means that even if we make considerable improvements to animal welfare, we may not be able to pass this on to consumers – we operate in a competitive market and cannot simply charge more for key products. They will shop elsewhere. For example, we have recently set up a series of farm trials in which we are attempting to greatly improve the welfare of meat chickens housed intensively indoors. We give them much more space than is common in Europe or the US and we give them straw bales as perches. The houses have windows that let in natural daylight. The chickens cost more to produce, but the marketing of this product has proved difficult. Customers still see a large shed with large numbers of birds inside and it is this that they dislike. In fact, the inside of a free-range shed, even with as few as 600 birds in it, is viewed unfavorably by many people. If it is the large number of birds together that causes people to believe they have poor welfare, then all our attempts to improve welfare in other ways will go unacknowledged. The best (the widespread view that all chickens should be outside) is in this case the enemy of the good (improving the welfare of the much larger numbers of birds that are still kept inside).

Even trying to sell the welfare benefit of keeping pigs with their tails on isn't easy because most consumers are not aware that most pigs have their tails removed. This means that a change that is considered to be a major welfare improvement to those who know about pigs and represents a change to pig farming of such a massive scale that most farmers will have nothing to do with it, causes not even a ripple of approval from most customers. Only if it is explained to them that most commercially reared pigs are tail-docked will they begin to appreciate the welfare benefits of the new system. And to do that, it is necessary to expose a welfare issue affecting the whole of the rest of the industry, which in turn will affect sales of "ordinary" products. It can be difficult to say how good one product is when it has an implication that other products on sale are to a lower standard – as a retailer (acting on behalf of our customers) we always chose to promote the positives of a system and the added advantages it has, rather than criticize other production systems.

Welfare improvements are therefore possible in two key ways, for a mainstream retailer: first, selling and promoting high welfare products which have simple consumer messages (e.g. free-range chickens, outdoor-bred pork) which customers will pay a premium for; and second, driving improvements in mainstream production systems through livestock codes of practice and pioneering systems, which are often much more subtle but have the ability to affect the lives of many more animals.

The achievement of that critical mass of farmers producing food in ways that allows us to insist on better standards of welfare because they and we know that the products will sell, can be promoted by us through innovators like FAI and our champion farmers. We can talk to our customers and understand their expectations and priorities. The animal welfare organizations have a major part to play, not only by educating people as to how their food is produced but also by rewarding producers and retailers who make genuine improvement in animal welfare. It is a public acknowledgment that things are changing for the better, even if they are not yet as the charities themselves would like them to be. Prizes and awards thus have a major impact on public opinion generally and help form part of the view that the public has on what retailers do. This in turn influences consumer loyalty, which in turn influences the way farmers see their own prospects if they took the plunge and ran their farms in a different way. No one foresaw the incredible growth in free-range eggs that has taken place in Europe over the last 5 years. Free-range eggs now account for over 50% of the UK market, clearly driven by public opinion but quite unexpected in its magnitude.

So, retailers have an important part to play in bringing about change in farming practices, both as recipients of changing demands from their customers but also, I believe, as active movers and shakers using commercial influence to encourage change that wouldn't otherwise have occurred as quickly. I have tried to show that the large retailers have a role at every stage of the process, from the initial research findings, to the receiving of the prize for achieving a particular improvement in animal welfare. But the process is complex and we still do not understand fully why some changes gather pace and become the norm, and others fizzle out and die, despite being apparently good ideas.

What is clear is that we are all in this together and that people who really want to improve animal welfare have to work as a team, even though they may variously find each other too extreme, not extreme enough, too fast, too slow, too commercially minded, not commercially minded enough, and so on. Scientists, for example, can often be frustrated by the fact that their research findings are not acted upon sooner by farmers, when they feel they have already demonstrated clear animal welfare benefits. But many farmers have no idea at all what research has been done, even when it has been sponsored by their own industry. Farmers have no access to the usual channels by which scientists communicate such as research journals and they often do not use the internet or have time to read the many leaflets sent through their post. The farming press is often not concerned with promoting innovative new systems. Agricultural shows are one way of reaching out to farmers but few scientists attend these. Here is one area where retailers could play a much more active part, by helping with the communication of the results of research so that farmers were at least more aware of what could be achieved even if, initially, they were sceptical that it could work in practice.

But it is in removing scepticism and promoting "rollout" that retailers can have the greatest effect. Links with organizations such as the Food Animal Initiative in the UK are the first step. As a fully commercial farm with a determination to put research into practice, FAI have given me the confidence to push for change even when I am being told by farmers that what I am suggesting is impossible. If I've seen it working and seen the improvements in animal welfare within the tough environment of UK commercial farming, then I know that change is possible. It may not be easy or straightforward and there may be specific pitfalls to be avoided and standards to be achieved, but I know I can go ahead to the next stage. I can take the second step of persuading champion or pioneer farmers to adopt a similar system. I need to make sure that they have the expert advice they need to make a success of what may be a new and risky business. As long as they know that they are going to have help if they run into problems, then I feel I can really make a difference to their chances of success and so to the success of others who will feel confident of following in their footsteps. As Temple Grandin puts it, it is all about achieving critical mass or tipping point and retailers are an essential part of making that happen.

BRINGING ABOUT CHANGE: A RETAILER'S VIEW

11 Animal welfare as a business priority

Keith Kenny *is a Senior Director for McDonald's Supply Chain in Europe. Keith leads the development and implementation of the company's sustainable supply strategy and food-related issues management. He pioneered the development of the pan-European McDonald's Agricultural Assurance Programme including McDonald's Animal Welfare Programme. Keith holds a BSc (Hons) in Food Science from Kings College London.*

Over the last 30–40 years, animal farming has become increasingly more industrialized and, at the same time, consumers have become increasingly removed from where their food comes from. Many people now have little idea about how their food is produced or what sort of processing it goes through. Some even find it difficult to make any connection at all between farms and the food they eat. However the ongoing list of food scares and negative publicity about food has made many people begin to realize how important it is to know much more about the different components of the food they eat. They want food that is not just safe but nutritious. They want to know what is in the food that they eat, where it comes from, and that it was produced under ethically acceptable conditions. In other words, they want transparency in how their food is produced and animal welfare is a big part of what they want to know about.

This increasing demand for transparency about the provenance of food continues to gather momentum and has become a key business priority for leading food companies. Corporate Social Responsibility forms part of the everyday language in the boardrooms of responsible companies all across the globe. In the past good food companies were concerned with nutrition, the environment, and animal welfare but it was not necessarily clear how some of these requirements fitted in with the overall business goals. Now that has all changed. Leading companies clearly understand that an important component of their success over the long term is to build trust in their brand and to convince their customers that they, as a company, can be relied on to deliver food that is good for them, good for the environment, and that animal products come from animals that have been well treated. Animal welfare thus becomes part of the long term image of that brand and results are measured in terms of the much longer term benefits of its contribution to increasing customer confidence and trust. If it is viewed like this, as an investment, not a cost, then the process of bringing about change in animal welfare is more practical and effective, especially since this model does not demand instant returns for any welfare improvements. I would like to share with you some of the ways in which McDonald's – the world's largest food service retailer – has been attempting to bring about improvements in animal welfare in Europe with this approach

Up until recently, McDonald's growth strategy was dominated by opening up new restaurants in new and existing countries. There are now 31,048 McDonald's restaurants in 118 countries across the world. In Europe there are over 6300 restaurants in 40 countries employing more than 270,000 staff. Although we are still opening many new restaurants, our main business strategy is to increase sales in the existing restaurants. We do this either by attracting new customers or persuading our existing customers to visit us more often. Either way, the trust of our customers in the McDonald's brand is essential. Because animal welfare is important to our customers, it is important to us too.

One of McDonald's greatest assets and one of the main reasons we have been able to develop a truly global restaurant system, is the strength and structure of our supply chain. It is important to understand the structure of our supply chain and the way we work with our supply partners, as this is fundamental to our ability to address some of the challenges of improving welfare.

The foundation of our supply chain system is partnering, our suppliers have grown with us in new and existing McDonald's markets, and they work hard to apply continuous improvements on our behalf. We involve our suppliers in our business, all the way through to customer delivery.

Trying to improve animal welfare, however, is not always a straightforward process. We don't own any farms. We don't breed or grow our own animals. We don't own any manufacturing facilities, transport networks, or abattoirs. We are thus completely dependent on large numbers of other people who constitute our supply chain. For example, for our beef patties sold in Europe, we use animals raised on about 500,000 different farms spread across the continent. Furthermore, the meat we buy from these half a million farmers has been largely raised for the dairy industry, with no direct connection to the beef supply chain until after their life as dairy cows is over. It is not easy to influence the welfare of these animals when they have nothing at all to do with McDonald's until after they are slaughtered, de-boned and converted into hamburgers, all by people who are not employed by us. Therefore, while McDonald's is interdependent with its supply system, we do not own it. We believe this approach delivers the safest, best quality food, produced according to the highest standards. However it does throw up challenges when trying to exercise positive changes in areas such as animal welfare a long way back up the supply chain.

However we have a commitment to improving the welfare and sustainability of the animals and farming practices used in our supply chain. We took the decision that the best way to improve the welfare of the animals in our supply chain was not to target individual farmers but to try to improve standards in the entire industry. We were aware that many countries already have their own farm assurance schemes (Chapters 12 and 13), which set standards for their farmers, covering many areas such as good agricultural practice, environment, some animal welfare, etc. Although we were not totally comfortable with the low level of some of the standards, we saw no point in setting up yet another assurance scheme with the McDonald's label and a whole new set of welfare requirements for farmers to meet. Rather, we devised an assurance *program* to try to influence and upgrade the existing farm assurance standards in each country. The McDonald's Agricultural Assurance Programme (MAAP) was started in 2002 with a set of documents that detailed what standards we would expect to see on our supply farms now and also a set of aspirational standards that we would hope to see adopted in the future. The six key policy areas in this program are:

- Environment
- Good agricultural practice
- Animal welfare
- Animal health
- Transparency
- Genetics.

We aim to work in partnership with as many farm assurance schemes as possible, so that when they update their standards they might incorporate some of our additional requirements. Since the best farmers already achieve many of these standards by incorporating them into assurance schemes, it ensures that the others are brought along too. We also prefer farm assurance schemes that are independently audited as this adds more weight to our consumer messaging and confidence.

The McDonald's requirements are updated each year and our intention is that today's aspirational ones become tomorrow's required ones and then new aspirational ones are added. That is our way of ensuring continuous improvement. We don't have to own an assurance schemes in order to have influence on the standards within it.

Our suppliers are also part of the process of deciding what our next set of requirements should be. This is done via our product-specific Technical Advisory Groups, which bring together our suppliers' experts, such as; vets, nutritionists, and farmers, to share solutions to common problems, such as the ban on the use of antibiotics, and agree practical and aspirational requirements for our annual review of the MAAP standards. The suppliers are very much part of this long term enterprise.

From these Technical Advisory Groups, we also get an idea of best practice, not just in one country but from across Europe. This means that if one country thinks that we are asking too much, we can point to other countries where the same thing is already happening. Here is a case where a large multi-national company can have positive influence across national borders.

We measure the degree of compliance of each farm assurance scheme used in our supply base against our MAAP requirement by performing a GAP analysis (Good Agricultural Practice; http://www.nri.org/NRET/SPCDR/Chapter4/agriculture-4-4.htm), so that we have an idea of how closely each scheme meets the standards we expect. Schemes that achieve a low degree of compliance are targeted for discussion. Indeed our suppliers will often approach and lobby existing farm assurance schemes on our behalf to incorporate our MAAP requirements into their standards.

Supplier compliance with our MAAP targets is measured annually and these results are feed into a much broader annual business review, looking at all aspects of the business from food quality to management strength, assured supply, and financial stability. These annual review results are used to reward the best performing suppliers with additional McDonald's business. In this way, compliance with our welfare standards can be directly related to supplier growth and new business opportunity.

Another approach we are excited about is our participation in The European Animal Welfare Platform. This platform brings together food retailers, food producers, farmers, animal welfare groups, and academia to discuss scientifically based improvements in animal welfare. For the first time in the area of the welfare of farmed animals, this project harnesses the efforts of principal stakeholders throughout the supply chain in order to address the growing societal need of consumers and citizens for a high welfare quality and increased transparency of production. The current European Animal Welfare Platform was initiated a couple of years ago and these key stakeholders have met regularly since the end of 2005.

We are, of course, aware that improvements in animal welfare can often cost money in the short term and this is yet another advantage of our partnership approach and long term relationship with our suppliers. We get a very clear idea of any of the extra costs involved and can discuss with them how they are to be managed.

This leads me on to an important difference between the food retailer business and the food service business such as ours. For a retailer, the commercial value of improved animal welfare is apparent from the choice its customers make. People can show that they value higher standards of animal welfare by buying high welfare products and showing that they are prepared to pay more for them. But when customers come into one of our restaurants, they do not have such a choice. We provide the food and they either eat it or go elsewhere. They do not show that they value animal welfare because they choose one of our products over another, but would show it by not coming to us anymore. So we as a company have to make the right choices. For example, 10 years ago, McDonald's UK started to serve only free-range eggs in all its restaurants, this was not because our customers demanded it. In fact, we had had no consumer pressure to move to free-range eggs. It was a company decision and it cost us considerable amounts of money to implement it. We did not put up the prices in our restaurants. We simply absorbed the costs because we felt that this was the right thing for the company to do. So this was an example of a cost that we incurred in the interests of what we saw as improved animal welfare. We saw it as an investment in the future through the trust our customers would have in our products generally. Now over 92% of all the whole eggs used in McDonald's Europe are noncaged and we are committed to get to 100%.

But this put us in a very curious position because hardly anybody knew about what we had done. Even now, most people are astonished to learn that McDonald's uses free-range eggs. We have perhaps in the past been guilty of not telling our animal welfare stories loudly enough. Maybe there was some concern that if we say it too loudly, then people tend not to believe us or, if we get one thing right, they will start criticizing us for not being perfect in every respect. We have started to address that now, but we feel that the best way of getting this story heard is though other people's endorsement of what we do. The recognition by the RSPCA and more recently the Good Egg award from Compassion in World Farming have been very helpful in this respect. We are really grateful that the welfare organizations have realized that carrots are often better than sticks and give credit where it is due. We definitely all need each other – business, animal charities, farmers, and consumers – if we are to achieve real change in standards of animal care.

That gives some idea of what we are currently doing for animal welfare and how we are pushing for change across the agricultural industry. Encouraging everyone to adopt the best practices that some farmers are already using is a powerful way of raising standards generally. However, we want to do more than just push for what is currently best practice. We also want to look further to the future and think about aspirational standards – ways of keeping animals that do even more for their welfare and go even further towards keeping animals in ways that both the animals themselves and our customers want. As I stressed earlier, McDonald's takes a very long term view of what "good business" is and does not expect immediate returns on everything it does. This means that we are prepared to invest in projects that may take some time to come to fruition but help us to see where animal farming may be going in the future.

One such project is the Food Animal Initiative near Oxford, UK, of which McDonald's is proud to be a founding sponsor and partner. This, as described elsewhere in this book, is a long term project set up to develop commercially robust alternative farming methods that significantly raise animal welfare standards as well as addressing human health and environmental concerns. Together, we are committed to develop new farming practices, whilst helping secure the future livelihoods of rural communities. The project is being run in conjunction with Oxford University's Zoology Department but has the great advantage from our point of view that it is a fully commercial enterprise. This means that the methods developed there are firmly rooted in commercial reality. They may be innovative but they are also practical and known to work in practice. Seeing new ideas at work at FAI therefore means that we can begin to introduce them into our Agricultural Assurance Programme, knowing that they work and make business sense. FAI gives us a glimpse into a new future for animal farming, one that we can then begin to make into reality. We see it first at FAI and then use the strength and vision of McDonald's to help shape the future we want.

Some of the work at FAI could be described as evolutionary – small innovative steps that make a large difference, such as planting trees for free-range chickens. These are important as they show how any farmers can make small, often inexpensive, modifications to the way they keep animals that will nevertheless improve animal welfare. Other projects, however, at least seem at the time to be more radical, more in the way of revolution. Keeping pigs without cutting off their tails is seen as revolutionary by many farmers who are convinced that it cannot be done without the risk of outbreaks of tail-biting. Keeping broiler breeders (the parent of meat chickens) outside as free-range birds is also seen as revolutionary by the many people who claim that free-range is such a risk to biosecurity that no breeding birds should be allowed outside. Seeing unmutilated pigs and thousands of healthy broilers gives us the confidence to believe that, despite what the pessimists say, these are feasible, viable ways of keeping animals that result in significant improvements to animal welfare, We can then take them forward and drive the process of making them more widely adopted.

However, being a customer-driven organization also means that McDonald's needs to stay ahead of its customer's expectations and look further into the future. Some of our customers have relatively moderate incomes and are not demanding that we improve animal welfare standards. It is we who look to the future and decide that, as a long term investment, we want to invest in food quality, of which animal welfare is a very important part. The McDonald's supply chain model of looking towards a long term supplier partnership used to be quite a unique approach, but now more and more companies are seeing the value of this way of thinking, especially in respect of social accountability issues. It helps that like-minded companies are able to take a longer view and incorporate animal welfare into their strategic plans.

So what does the future hold? Where might we be going? It is already possible to identify some areas where change is necessary. We need better and more robust ways of auditing the welfare of farm animals so that we can ensure that the standards we want are maintained and, more pertinently, that they are effective. The process of auditing welfare standards needs to be made much easier so that it can be more easily applied. One route will almost inevitably be to refine the "outcome" measures at slaughterhouses and on farms, so that we know what has been happening to the animals during their lifetimes. For example, the incidence of ulcers on the feet of chickens at slaughterhouses can be used to give an indication of the condition of the litter the chickens were standing on when they were alive. Damp, messy litter gives rise to lesions on the feet (pododermatitis), which is thus a "telltale" sign later on. We may be able to require farmers to achieve low target values of foot ulcers as a way of monitoring the conditions inside their houses.

Dr Temple Grandin showed how targets that she called Critical Control Points could drastically improve standards in slaughterhouses. She showed that if slaughterhouses were set targets – such as a specified low number of cattle bellowing or slipping over – the numbers of cattle slipping over dropped dramatically. She showed slaughterhouses what they could do to stop their cattle slipping over – simple things like improving the floor or removing objects that frightened the cattle. She then told the slaughterhouse to achieve their Critical Control Point targets, which they then did. The next step is to move beyond these Critical Control Points (achieving very low levels of abuse) to achieving high levels of positive welfare, which is why we need robust, easy to use audit measures. We need species-specific welfare measures that have been validated for all species from dairy cows to chickens.

Looking further to the future becomes more difficult as the world is changing rapidly, and even in 10 years time there may be such changes that our current speculation will be wide of the mark. For example, changes in the cost of fuel could radically change the relative "carbon footprints" of food grown in different parts of the world. If chickens reared in Europe are fed with soya grown in Brazil, then there could be an argument for rearing chickens in Brazil not Europe. In many ways, the Brazilian climate is better for free-range chickens than cold, damp, European winters. The ancestors of chickens (jungle fowl) were, after all, tropical birds and if feed, heating and labor costs are lower in Brazil and other South American or Asian countries, then arguments based on the impact on global warming could change our ideas of where they should be kept. On the other hand, political consideration may put food security and self-sufficiency at a premium. Technological advances may give us new sources of food so that we won't need

ANIMAL WELFARE AS A BUSINESS PRIORITY

soya to feed our animals or even, in the longer term, provide a substitute for animal products altogether. The impact of biofuels on agricultural land and what this might do to the human food supply has only just begun to impact on our ways of thinking.

We don't know what the very long term future holds and we need to keep an open mind. We do know that there is rising demand for meat worldwide that is likely to continue for some time. At McDonald's we want to meet this rising demand for animal products in ways that put more emphasis on how the animals are reared and cared for. We want to move farming in the direction of giving our customers what they want in terms of quality of their food and giving animals what they want in terms of quality of their life. We also believe that this makes good business sense.

Commentary

If consumers are to make informed choices about the food they buy, they want to be confident that they can believe what they read on the label. If meat is described as "free-range," they want a guarantee that it really is. If a supermarket claims that their produce comes from animals treated in a welfare-friendly way, they want to know what that means and also that someone is making sure that the claimed standards are actually being delivered. In other words, for high welfare farming to succeed, there need to be systems of monitoring and policing, or, to put it in a more upbeat way, positive approval for those farmers who can demonstrate that they have achieved certain standards. This monitoring and approving generally comes under the heading of "farm assurance" and the bodies that do the monitoring and approving are now many and various. Some are run by groups of farmers, others by animal welfare charities. There have been attempts to set standards that apply to whole countries, such as Canada (www.inspection.gc.ca), the United States (awic.nat.usda.gov), to groups of countries such as the European Union, Australia (animalwelfare@daff.gov.au), and Zealand. With this diversity of schemes and standards now in operation, we asked two people who have been actively involved in developing and implementing such schemes, David Main and Sir Colin Spedding, to give us their views of the issue of how to ensure that people are actually able to buy the farm produce they believe they are buying.

12 Providing assurance on welfare

David Main *is the BVA Animal Welfare Foundation Senior Lecturer in Animal Welfare at the University of Bristol. He is a RCVS recognized specialist in Animal Welfare Science, Ethics and Law. His consultancy work has included providing advice on retailer standard. He is also a member of the UK Farm Animal Welfare Council.*

Providing a genuine assurance to your customer on the quality of your product is obviously a key part of any successful business. Since knowledge of husbandry conditions of farm animals may influence the purchasing behavior of consumers, providing assurances on welfare has been included as a quality attribute of livestock products alongside physical attributes (such as taste) and other ethical concerns (such as environmental issues). Many large retailers and catering companies have recognized this moral concern and have consequently developed (often complex) strategies of their own to either promote or defend the welfare standards of their products. This chapter will review how voluntary certification schemes are being used to respond to this moral concern.

What are the approaches developed by and for food industry?

Voluntary schemes, in which a certificate is issued for achieving certain standards, have been developed by industry to provide relevant assurances to their consumers. The standards may vary in the level of welfare required and in their scope, i.e. which types of issues are covered. For example in the UK, the larger national assurance schemes cover relevant legislation and codes of practice for food safety and environmental issues as well as animal welfare. Schemes such as the RSPCA Freedom Food standards, on the other hand, focus primarily on animal welfare but insist on a higher level than legislation. Organic certification schemes aim to cover a wide range of animal welfare and environmental issues, also to a higher level.

As discussed more fully in the next chapter, a core foundation for animal welfare assurance in the UK has for some time been national farm assurance schemes such as Assured Food Standards (www.redtractor.org.uk). The principle behind these large scale assurance schemes is that there should be standards to address minimum legal requirements and elements of good practice in three core areas (food safety, animal welfare, and environment). The level of the standard is usually pitched to enable the vast majority of producers to participate.

Another strategy available in the food industry is to use welfare-focused schemes or initiatives to generate welfare-related claims on their products and to insist on higher standards of welfare. The RSPCA (Royal Society for the Prevention of Cruelty to Animals) Freedom Food scheme is an example of such a scheme designed to achieve high standards of animal welfare (Royal Society for the Prevention of Cruelty to Animals, 2001) on farm, during transit, and at slaughter. The Freedom Food scheme requires members to adhere to welfare standards set by the RSPCA in association with species-specific technical working groups. These groups include producers, industry experts, veterinary surgeons, and animal welfare scientists. The RSPCA has produced standards for many species including cattle (dairy and beef), sheep, poultry (laying hens, broilers, turkeys, and ducks), pigs, and salmon.

Development of these voluntary schemes has historically been driven by the need to provide assurance on food safety. In the UK this was driven by various food scares such as BSE, salmonella, and E. coli 0157. This resulted in the Food Safety Act (1990) which requires all food retailers to demonstrate due diligence in ensuring that the food is safe to eat, i.e. carry out "all such checks of the food in question as were reasonable in all the circumstances." A critical benefit of developing these national schemes was avoiding the costly need for each retailer to undertake their own checks on each farm that might enter their supply chain. Therefore, even though these schemes are voluntary, all producers wishing to sell to the larger retailers would be required to be a member of such a scheme.

The development of a large scale certification system has provided a vehicle for defining and verifying producer compliance with other issues such as animal welfare and environmental issues. The key attributes for success in this process has been the industry wide consensus of the necessary standards plus independent inspection and certification by certification bodies accredited to EN45011 standards. Competition between certification bodies and development of schemes able to certify more than one type of farm business (e.g. beef and cereals) has lead to very cost-effective certification at farm level.

The influence that retailers have on animal welfare standards is a key feature of the livestock industry in many countries and this is a recurrent theme of this book. A key mechanism for this influence is the interaction between the larger retailers and the national farm assurance schemes. Retailers exert their influence over these schemes by creating demand for certification by requiring their suppliers (e.g. abattoirs and dairy companies) to only purchase livestock or milk from farms certified to the minimum standards and by direct negotiation with the standards setting body or by asking the certification bodies to assess compliance with additional requirements during their normal certification visits.

Do the schemes deliver a genuine welfare assurance?

Any credible welfare-related inspection system that is in addition to a normal government-controlled enforcement system is likely to have some impact on compliance with welfare legislation. Provided standards include the relevant legislation, the assessor is adequately trained, and the reporting system is thorough then a certification system should raise awareness and compliance of the legislation requirements. In the UK, the proportion of production within the Assured Food standard scheme ranged from 65% in the sheep sector to 95% in the dairy sector (Department for Environment, Food and Rural Affairs, 2007a), whereas over the same period (2005–6) only 3834 official government welfare inspections (i.e. less than 10% of holdings) were conducted (Department for Environment, Food and Rural Affairs, 2006). Simply by their presence on the farm and their role in reporting noncompliance with standards that include welfare legislation one would expect these schemes have an effect beyond existing enforcement systems.

One simple method of demonstrating this is to examine the noncompliance rates of farms on their first assessment compared with subsequent (often annual) surveillance visit, i.e. before and after membership approval. Evaluation of the noncompliances generated by one assurance scheme in the UK, Farm Assured British Pigs, demonstrated that 67% of the 372 noncompliances examined were welfare-related (Main and Green, 2000). Noncompliance rates identified by government veterinary surgeons conducting the inspections in this voluntary scheme fell from 16.6% on initial visit to 6.9% on subsequent surveillance visits with significantly lower noncompliance rates for medication and movement records, potential injurious structural defects, provision of alarms and ventilation failure protection, and provision of hospital pens.

However, such analysis of noncompliance with welfare legislation, whilst intended to benefit the welfare of the pigs, may not necessarily result in an improvement in welfare outcomes. The RSPCA commissioned the University of Bristol to evaluate the animal welfare impact of the RSPCA Freedom Food (FF) scheme on dairy cattle, pigs, and laying hens. For this evaluation the most important elements of "welfare state" were first established through an iterative review of expert opinion and then incorporated into a systematic program of observations and records-related welfare measures (e.g. production, reproduction, disease, and behavior). The full description of these assessments in dairy cattle, pigs, and laying hens are beyond the scope of this book (Main et al., 2003; Whay et al., 2007a, 2007b). However, an interesting example was the finding that lameness and lying area discomfort in dairy cattle were found to be at high levels in FF as well as non-FF farms. These priorities were identified by evaluating the number of FF farms above "intervention levels" that had been derived from another consultation with experts. The farms above these intervention levels did not reflect the extent of compliance with the FF standard. The FF standards are largely resource-based (e.g. housing conditions) and the University of Bristol assessment was animal-based (i.e. behavior, physical condition, and health records). It was, therefore, very possible that a farm was above the University of Bristol intervention levels for one or more welfare measure but still fully compliant with the FF standards. An important conclusion to be drawn from these studies was that welfare problems and priorities for action are specific to individual farms, highlighting the importance of requiring farms to actively manage health problems and for certification schemes to incorporate outcome-based welfare assessments.

What are the political influences on these market schemes?

Farm assurance schemes are voluntary certification schemes where producers can choose to join the scheme and abide by the scheme rules/standards. Even though the major UK national schemes have at times received financial support from the government, they are primarily private sector initiatives and are usually owned and driven by the relevant industry bodies. The potential influence of these schemes on the farming industry, however, has been recognized by several bodies and is becomingly increasingly the subject of political influence and recommendations from various reports. For example, the Farm Animal Welfare Council (2005) examined the welfare impact of farm assurance schemes and recommended that "... scheme owners should work towards refining their standards and inspection procedures to achieve an increasing inclusion of welfare outcomes ... ". This is also further supported in the welfare labeling report in 2006 that concluded that "welfare labelling should take into account the welfare of individual animals over the entire course of their lives (including pre-birth, on-farm, during transport, at markets, and at slaughter). It should be based predominantly on welfare outcomes (i.e. the measurable welfare status of the animals involved in producing the product) rather than on other indicators such as the production system." (Farm Animal Welfare Council, 2006.)

The importance of welfare indicators has also been recognized at a European level. In the Community Action Plan on the Protection and Welfare of Animals

2006–2010 produced by the European Commission a clear goal was to introduce standardized animal welfare indicators that could classify the hierarchy of welfare standards applied (from minimum to higher standards) in order to assist the development of improved animal welfare production and husbandry methods and to facilitate their application at EU and international levels. These standardized indicators are being produced by a large European research project, Welfare Quality (Blokhuis et al., 2003).

Differences between consumer expectations and scientific approaches to welfare

Evaluation of the differences between the consumer expectations and the approach to welfare assessment advocated by scientists has been an important area of investigation. The Welfare Quality project has identified some critical tensions between these groups (Vessier and Evans, 2007). First, many welfare scientists have advocated that an animal welfare assessment should be based on primarily outcomes that indicate the state of the animal such as physical and mental health. These measures are independent of the husbandry system and, thereby, allow comparisons between systems. However, consumers also often wish to know about the environmental conditions, housing systems, and other resources that are provided for the animal. Second, the value of " naturalness," which has long been debated amongst welfare scientists, appears to be strongly valued by consumers, even if it merely involves the aesthetic appearance of animals in green fields rather than a more animal-centered approach considering the animal's performance of natural behaviors. Third, scientists have taken the approach of assessing several parameters in scientific investigations as many aspects of an animal's life appear to be valued by the animal. This is reflected in the different components of the Five Freedoms (Farm Animal Welfare Council, 1993). However, many consumers seem comfortable with thinking of welfare as a single holistic concept rather than a combination of several different components. Finally, consumers seem to find it difficult to separate welfare considerations from other societal concerns such as environmental sustainability, food quality, and human health.

Addressing these differences between consumer and scientist expectations of a product information system is important for the industries involved. A relevant example of such conflicts would be the purchasing preferences shown by consumers for eggs labeled according to their system of production. Whilst it may be possible to devise a purely animal-based outcome measure to summarize the animal welfare state of chickens from one particular unit, the current European labeling legislation requires that eggs are described according to production system such as caged, barn, or free-range. UK Consumers have demonstrated a strong preference for eggs from noncage system. In 1998, free-range eggs made up around 15% of total egg throughput and in 2005 this had doubled to 30% (Department for Environment, Food and Rural Affairs, 2007b). Since the husbandry system has such a significant impact on the birds' ability to perform natural behaviors, such a husbandry system label yields useful welfare information. However, in an ideal world the welfare scientists might like to include information on other welfare relevant indicators in addition to behavioral freedom, such as levels of injurious pecking, fractures and mortality, which can vary very significantly in noncage systems.

An overriding assumption underlying the call for welfare labeling is that consumers value information about welfare provenance of the products. This is supported by some surveys, for example the Eurobarometer survey (European Commission, 2005) asked EU citizens in 25 countries "Would you like to be more informed about the conditions under which animals are farmed in your country?" 39% said yes probably, 19% said yes certainly, whereas 26% said probably no and 13% said no, certainly not. Even though a majority were positive it is interesting to note the relative small proportion of the population that were strongly in favor of this information. Other information from focus group work is again not conclusive in support of a complicated information system. Mayfield et al. (2007) after a series of work in focus groups concluded that "although most people were concerned how animals are treated, for many, there was an inbuilt disassociation between meat and animals, and a definite resistance to having to think about how animals are reared" and "there was a consensus that ... they would in fact rather not have the choice and thus be forced to do the right thing." The provision of additional information was seen by some participants as time consuming and they preferred to be provided with something simple they could recognize and trust.

What is the future for farm assurance?

As discussed an increased reliance on assessment of welfare outcomes has been advocated as important for the future development of schemes. In addition to improving the assurance provided by the scheme, assessment of these measures would be useful information to the producer and could be included in benchmarking reports alongside productivity data. Whilst provision of advice would not be possible for an accredited scheme, farm assurance schemes can collate production or animal welfare performance data during their visits and provide benchmarked reports. This would enable the producer to identify their unit's relative strengths and weaknesses and monitor the impact of husbandry changes. The schemes are in a unique position to provide such services as they are credible, independent, and visiting a large number of similar units.

Another way in which assurance schemes could progress is by facilitating more detailed differentiation of products based on different levels of production standard. As previously discussed, it is clear that many consumers value a single common minimum standard, however, some consumers do value differentiation of various features of production system such as feeding, housing, and location. For small scale producers wishing to provide this information to consumers, using the services of the existing certification bodies would seem cost efficient as they could also certify the farm to the relevant national farm assurance scheme. Some certification bodies already offer assessment against more than one standard during the same visit (e.g. organic and industry scheme). This could be further extended during the same visit, either based on outcome-based assessment systems such as welfare quality or husbandry system descriptions such as outdoor or free-range. Another feature of certifying farms against more than one standard is the flexibility in the final destination of the product. For some meat producers not all of the carcase can be used for a premium labeled product such as organic, so a dual certification means that lower value cuts can be used in the conventional sector.

As schemes mature and experience grows the level of assurance provided is likely to increase. The accreditation to EN45011 (ISO Guide 65) has clearly had a significant impact on inspection and certification in the UK. A key feature of accreditation to this standard is the internal quality management systems. For example, all assessors are monitored regularly and internal audit of the whole process ensures that any deficiencies in the system are corrected in a timely manner. This self-evaluation, verified by the national accreditation body, should promote continuous improvement in the certification process. In addition to these formal systems the continuous media scrutiny of food safety and animal welfare issues should ensure that standards will not slip in the foreseeable future.

The standards used for certification schemes are already used on an international basis as retailers are aware of the importance of the standard of food they sell whatever the origin. There is likely to be further formalization of these international standards, for example, led by bodies such as OIE (Orgnaizacion Mundial de Sanidad Animal (World Organization for Animal Health) http://www.oic.int/eng/en_index.htm). For animal welfare, the assessment systems generated by the Welfare Quality project aim to support such international agreements. The relative importance of other issues may well also change in the future. For example, environmental impact is likely to become much more important for all livestock products. Similarly protecting the working conditions of those involved in food production, such as in the Fairtrade certification scheme, may become more important for livestock products.

In conclusion, farm assurance is here to stay. Credible, independent voluntary certification schemes are addressing animal welfare at various levels and sometimes in combination with food safety and environmental issues. Retailers have had a critical influence in creating the need and shaping the content of the schemes. This critical requirement to demonstrate a credible transparent assurance to their consumers will remain for the foreseeable future. As discussed by Colin Spedding, this has led to the formation of a well organized and cost efficient infrastructure in the UK. However, there are opportunities for better use of this infrastructure, such as by improving the welfare assurance systems and by offering more differentiation of products according to internationally agreed standards.

13 The role of assurance schemes and public pressures

Professor Sir Colin Spedding CBE *retired as Pro-Vice-Chancellor, Professor of Agricultural Systems and Director of the Centre for Agricultural Strategy (CAS) in 1990 from the University of Reading. He chaired the Board of the Science Council from 1994 to 2000, the Farm Animal Welfare Council (FAWC) from 1988 to 1998, the UK Register of Organic Food Standards (UKROFS) from 1987 to 1999, and the Apple and Pear Research Council (APRC) from 1989 to 1997. He is a past President of the British Society of Animal Science (BSAS) and the Institute of Biology.*

He is currently Chairman of Assured Chicken Production and on the Council of Management of the People's Dispensary for Sick Animals (PDSA), Special Scientific Adviser to the World Society for the Protection of Animals (WSPA), Adviser to the Companion Animal Welfare Council (CAWC), Chairman of the Farm Animal Welfare Trust (FAWT), and Vice President of the Royal Society for the Prevention of Cruelty to Animals (RSPCA).

As described by David Main, food assurance schemes are of two kinds. Those that cover all aspects of food production – food safety and environmental protection as well as animal welfare – co-exist with more specific animal welfare schemes. The food assurance schemes in the UK cover the bulk of agricultural production, with up to 90% of production coming from such schemes. They thus provide reassurance to the consumer that basic standards have been met. They also provide a means for bringing about change in large sectors of the industry, as large numbers of producers will all agree to change together. On the other hand, the more specialized schemes, such as the RSPCA Freedom Foods scheme, affect a relatively small percentage of production since relatively few farmers are members. Nevertheless, they may have a much greater impact than their percentage of the market suggests. They have a valuable role as pioneers, exploring practical improvements that may prove suitable for wider adoption at some time in the future. Both sorts of scheme are therefore important in their different ways in driving change in animal welfare.

It is worth noting, however, that most assurance schemes, of both general and specific sorts, tend to specify only the way something is produced, not the value of the product itself. For example, organic farming in Europe now has to conform to a Regulation (EEC No. 2092/91 of June 24, 1991). This sets down what a farmer has to do to be able to count his or her produce as "organic." All aspect of production have to be adhered to and inspected, but the Regulation makes no claims at all about whether the products themselves are in any sense better than those produced in other ways.

It is therefore a matter of personal judgment whether organically produced products are, in any sense, "better" than those produced by conventional agriculture. If you believe that the specified methods of production, processing, and distribution result in better quality (taste, food safety, nutritional value, freshness, etc.), that is your judgment: formally, no such claims are made or tested. Organic producers and organizations *do* believe all these things (not true – some do, but not all – but many organic consumers do believe and are prepared to pay which is why it has become a driver in the UK, and increasingly the global, market) and will claim them to be so, but this is not inherent in the Regulation. Animal welfare is somewhat different, since it is not a physical quality of the product – though treatment of the animal may affect product quality. However, many consumers would prefer to purchase products derived from "welfare-friendly" systems and these may be enshrined in the standards. So such claims are legitimate, provided that such systems really are better from an animal welfare point of view.

Food assurance schemes need to be carefully drawn up and implemented. In order to work, they need to be discussed with a variety of people and standards set that are neither too onerous for producers to implement or so weak that consumers have no confidence in them. As Chairman of one of the largest groups within the UK Assured Food Standards (Assured Chicken Production), I would like to describe some of the experiences we have had in setting up and running assurance schemes and the opportunities they provide for improving standards of animal welfare.

The main stream of farm assurance, however, is provided by Assured Food Standards which now represents the bulk of farmers. It covers all the main products (except eggs, see below) and in many cases up to 90% of the production comes from scheme members.

Assured Food Standards

AFS is run by the farmers who participate in it and who have come together to set standards. Technically, it is owned by farmers' organizations such as the National Farmers Union, the Food and Drink Consortium, with government representatives as observers. There are specific schemes covering all the main agricultural products (except eggs, which are dealt with by a similar Lion scheme). There are schemes for (meat) chickens, beef, lamb, dairy products, pigs, and "combinable" crops. As well as farmers' representatives, the AFS Board has five independent members representing food technology, consumers, and animal welfare organizations, as well as five members representing the food processing side of the industry.

All members are inspected, mostly annually, by independent auditors and are suspended if they fail consistently to meet the Standards set by each Scheme. As you may imagine, ensuring the competence, consistency, and impartiality of the inspection and certification process is a major task and so AFS operates according to international standards (ISO Guide 65) and national standards such as EN45011 by UKAS (UK Accreditation Service; www.ukas.com). It is interesting to note that whilst AFS schemes elect voluntarily to use EN45011, the same standard is now also prescribed for a similar purpose in several pieces of EU food legislation, notably in the regulations for organic foods. Currently, some 160 inspectors working for any one of half a dozen UK certification bodies carry out approximately 60,000 farm inspections every year against AFS standards.

Products from assured producers are entitled to bear the AFS logo (the Red Tractor), which is now on foodstuffs worth over £5bn. Food processors and packers must be licensed by AFS before they can use the logo. AFS is conscious of the fact that animal welfare and food safety can be compromised before the farm (e.g. animal feed, breeding stock) or after the farm (abattoirs, food processing, livestock transport, etc.). So it will only allow the Red Tractor logo to be used if the product has been handled or produced to the correct standards at every significant link in the chain. AFS has developed its own standards for some activities, such as transport of red meat animals and livestock markets. In other instances, it cross references other standards established within the industry, for example the AIC (Agricultural Industries Confederation; http://www.agindustries.org.uk/) feed schemes or the BRC (British Retail Consortium; www.brc.org.uk) Global Standard for food processing.

The beneficiaries

The main beneficiaries are the ultimate consumers of foods, who can be assured that the products they buy come only from certified producers who are independently inspected and suspended if they fail to meet all the Standards.Whilst the Red Tractor logo is primarily about the technical standards of food production, it now also includes an indication of the origin of the product. By the time the logo was introduced it was evident that this was a point of confusion for many consumers and there was distrust of some labels appearing at the time. Did a label "Made in Britain" on a pack of bacon mean that the pigs were British? Or that the

THE ROLE OF ASSURANCE SCHEMES AND PUBLIC PRESSURES

pigmeat was imported and cured in Britain? Or that the bacon was produced overseas and simply packed in the UK? Research from around the EU has indicated that in many countries there is an interest in knowing where food originated. And about the same time the Food Standards Agency (FSA) in the UK issued guidelines to provide more clarity in origin labeling. However these were just that, guidelines, with no specific force of law to ensure that they were followed.

So when the Red Tractor logo was redesigned in 2005 the opportunity was taken to include a very clear secondary statement of provenance in the form of a flag incorporated in the logo. The licensing conditions for the use of the Red Tractor logo are based closely on the FSA guidelines. They stipulate that the flag in the logo must indicate the place where the raw materials originated, that is, where it was farmed or grown. The scheme rules also insist that food packers and processors who are using the logo must have traceability and record systems that will validate that any product labeled with the Red Tractor logo was eligible to do so. And AFS has an integrity program that challenges those systems to make sure that they are in place and effective. Generally the Red Tractor logo is seen with the Union Flag showing that the product originates from UK farms. AFS technical standards are used overseas from time to time by producers or retailers who wish to demonstrate that they apply the same standards to the products that they sell irrespective of where they are produced. AFS ensures that standards must be used in exactly the same way with the same interpretations.

Retailers benefit from the Red Tractor initiative in that they can meet the aspirations of their customers in a way that ensures claims can be supported by evidence, and reference to transparent, independent Standards allows anyone to see exactly what is being claimed. These are spelled out in great detail, covering hundreds of separate points including feeding, housing, husbandry, medicines, and veterinary treatment. The Broiler Chicken Standards, for example, deal separately with hatcheries, breeding stock, meat chicken production, free-range, poussin, organic production, transport, and slaughter.

On the other hand, the benefit to farmers is often less clear. They receive no premium prices and may incur extra costs, they have to pay to become (and remain) members of the schemes, and they also have to pay the costs of inspection. Their economic benefits come from the fact that AFS-registered farmers have an extra market that would otherwise not be open to them. They can supply retailers who will only buy from registered suppliers.

Another benefit to farmers is longer term. Memories are often short, but "food scares" can be very bad for business. In serious cases, confidence can be lost in the whole commodity sector, not just in one supplier or group of suppliers that happen to be implicated. So an industry self-regulatory initiative is also important in maintaining the reputation of the industry. Preventing or at least minimizing media scares is an essential purpose of assurance schemes that deliver hidden benefits for farmers. From time to time we can even turn this into a more tangible benefit. So in early 2006 when the media was pre-occupied with threats from avian influenza, the UK broiler industry was able to point to the biosecurity precautions of the industry assurance scheme. And how did the consumer find chickens that came with that reassurance? By looking for the Red Tractor logo on the supermarket shelf. *Only* those producers who meet the standards can use the logo. The evidence was that UK consumers did retain confidence in chicken meat at a time when their neighbors across Europe were shunning it.

Improving standards

The main feature of the AFS Schemes is that they cover a very high proportion of the industry. Any advance in welfare standards therefore moves, in most cases, the whole industry forward. In this they are unique and represent the main opportunity to transform farm practice across all products and all producers. This is only possible, however, if the Standards are changed at a rate at which the industry can reasonably be expected to adopt them. Trying to move too fast would simply result in fewer members and less influence. Some might say that this is just an excuse for not moving standards forward, but the more general view is that it is a realistic position in the context of a global market in food supply. It is therefore vital to establish what improvements are practicable and economic: both are necessary to ensure wide adoption.

What change, exactly?

The change required in livestock production takes one of four forms:

(a)   more profit

(b)   better animal welfare

(c)   better environmental outcomes, or

(d)   better food safety

(a)   More profit

Everyone recognizes that profit depends upon costs and prices, some elements of which may be both unpredictable and uncontrollable by the farmer, and the efficiency of operation. Farmers thus tend to concentrate on the last of these, but what changes will improve efficiency?

THE ROLE OF ASSURANCE SCHEMES AND PUBLIC PRESSURES

Efficiency is essentially a ratio of output over input but even for a single enterprise there are several physical outputs (e.g. meat, wool/hides, milk, breeding stock, manure), some of which may incur a disposal cost, and a very large number of inputs (notably feed, rent, fertilizer, labour, housing, machinery, fuel, and veterinary fees).

Very often changes in an output produce or require changes in an input and both are ultimately expressed in monetary terms. So the effect of changes in one component of a system is difficult to predict and may vary from farm to farm and year to year.

Of course forward-thinking producers will be aware of the opportunity to break away from the constraints of "commodity price" and seek "added value." In economic terms it might be legitimate consciously to adopt a system that has higher costs provided that it delivers attributes for which the market is prepared to deliver a higher return. Some argue that this principle will apply to animal welfare conditions. "Higher" welfare systems may add to the costs of production but the producer will be rewarded by a higher return.

In reality, you only "add value" if the consumer values what you add. And the evidence at this time is that only a small niche segment of the market does in fact value higher welfare standards for animal products to the extent that they will positively seek them out and pay a premium for them. One key challenge is actually to understand the range of consumers' preferences in this area. The dilemma here is that surveys of consumer attitude are frequently quoted that do not correspond to what they purchase. We must be careful to separate what consumers say they will do when prompted, from what they say they will do when unprompted. And to separate even the latter from how they actually behave in practice.

(b)   Better animal welfare

All of the issues discussed under (a) will apply to the improvement of farm animal welfare, with the added difficulty that such improvements may be hard to measure. Incidence of disease, or lameness, for example, are relatively straightforward, but positive welfare (e.g. improved "happiness," quality of life, well-being, including mental) is more difficult to assess. So, here, too, changes in practices are seen as the means of improving outcomes, but can we be sure that they will, at what cost and with what other outcomes?

(c)   Better environmental outcomes

Environmental protection (of existing features) is easier to assess and cost: environmental enhancement is more difficult.

In all cases, we need to be clear as to who should change what and on what time scale.

Those who encourage, or even *force*, further change on farmers whether from government, retailers, pressure groups, or the public, need to be very clear about what changes they want and the costs and benefits of those changes before exerting pressure to change.

(d)   Better food safety

Better food safety is sometimes both specific and clear.

The big reduction in the incidence of *Salmonella* in broiler chicken production is a good example of what can be achieved by setting clear targets. In this case, everyone benefits. But there are times when there are conflicts between different targets.

For example, it might be generally accepted that "free-range" chickens have better welfare conditions than housed poultry. But in late 2005 and early 2006 the very serious threat of Avian Influenza suggested that only indoor production provided the necessary biosecurity to protect the health of both the poultry and the threat of human infection. (I would have said surveillance and rapid response have proved to be the main protection.) Another example of potential conflict is the prophylactic use of antibiotics, which might have attractions in preventing animal infection and suffering, but it is well established that this can lead to the development of antibiotic resistance in pathogens that could have grave consequences for the therapeutic treatment of both man and animals. These conflicts will obviously lead to very difficult decisions in order to reach a compromise position that is optimum for safety, welfare, environment, and commercial practicability.

Public pressure

The object of Assurance Schemes is to reassure the public that they can purchase identified products with confidence in the way they have been produced. The public can then exert enormous pressure by virtue of their purchasing power. If products are not bought, they will not be produced. Even though different qualities matter to different consumers, there are some over-riding qualities, such as food safety relating to whole classes of food (e.g. beef), that seem to matter to everyone. If the public comes to believe that they are unsafe, sales may drop precipitously – especially if the food is destined for very young children. Unfortunately, however, unfounded food scares can have just as big an effect as genuine concerns and reassurance by government – even by its independent agencies (such as the Food Standards Agency) – may not be effective, at least in the short term. That is why public education matters so much.

But people have many other things on their minds besides animal welfare and cannot be expected to be knowledgeable about everything on which they have to make decisions. There is indeed some truth in the proposition that democracy depends, to a large extent, on all of us having strong views about a range of subjects about which we are almost wholly ignorant. So it may generally be the case that most people have to trust the judgments of others: but who?

Big commercial enterprises are often suspect because "they only exist for profit" or "are only interested in making money for their shareholders." Other, non-profit-making organizations sometimes depend on frightening us in order to maintain their subscriptions/donations, which are paid by those who "want something done about it."

Curiously, the food retailers, who include some very large, profit-making organizations with enormous buying power, are amongst the most trusted by consumers. Why is this? It is, of course, partly skilful PR and good marketing and, although supermarkets claim to know what people want, they actually only know what people buy (and are persuaded to buy!). Behind all this, however, is trust. People believe those they trust: and large numbers trust the supermarkets.

The important safety element here is that trust is such a precious asset, hard to earn and easy to lose, that it is jealously guarded. If a retailer gets it wrong, trust evaporates, at great cost, and is hard to recover. This means that the major retailers, who can, and do, stipulate standards to which their suppliers must conform, try to genuinely reflect public opinion and, often, try to supply what they believe the public will demand in the future, especially in relation to animal welfare. Public opinion is therefore of great importance but it may be hard for the general public to decide exactly what standards they should demand when many people have little idea about how their food is produced in the first place.

The best hope for establishing trust in consumers is by farmers and processors following standards that are set and inspected independently by an organization that has no other purpose than to achieve and improve them. Assurance Schemes are the main vehicle for the improvement of obligatory standards and the consumer can support them by purchasing products carrying the appropriate logo. Such schemes have made enormous progress in the last few years, in relation to fresh and processed products, but there is some way to go in the food service industry, in catering establishments, and products containing mixtures of products, since consumers are much more likely to ask about the source of products at retailers than they are in restaurants. But some form of assurance, extended and strengthened in the future to cover all aspects of food production and sale, is the only way in which the consumer can exert real pressure for change.

14 Improving the welfare of cattle: practical experience in Brazil

Mateus Paranhos da Costa *has a background in animal science, with 21 years experience as teacher and researcher at São Paulo State University, at Jaboticabal-SP, Brazil. He has a particular interest in the welfare of dairy and beef cattle and in looking for practical ways to improve cattle welfare during handling. His career started at São Paulo State University in 1986, where he worked on farm animal behavior and welfare. From 1991 to 1995, he studied for a PhD in Psychobiology and spent 1991 in Cambridge University in the UK. He has published over 100 scientific papers and articles.*

Introduction

Efficient animal production is based on progress in genetics, nutrition, and management (including the development of facilities and equipment), and the resulting intensification of production systems. From the human point of view, there is no doubt that these strategies have provided economic and social gains. However, they have also brought side effects, causing environmental damage and impoverishment of animal welfare. We believe that it is possible to develop new strategies in animal production that ensure good productivity and high quality products, without putting the environment and the welfare of animals at risk.

In order to do this, we need to extend our knowledge about the biology of farm animals and to define ethical boundaries to clarify which procedures should be banned and which should be recommended. This is not an easy task. We need to establish a new paradigm for animal production so that as well as spending time and attention on the development of new techniques, we also consider the principles of animal welfare science. We have to be committed to the promotion of human and animal welfare, while ensuring environmental sustainability, satisfaction to the consumer, and profitability to the producers. The aim of this article is to present some positive (and practical) results achieved with beef and dairy cattle in Brazil, resulting from the development and application of best practices of cattle handling.

Difficulties in daily handling routine of cattle

In many paintings and pictures, the landscape of a cattle farm is idyllic, with cattle grazing peacefully on green pastures. This view does not reveal the reality of the daily routine of a cattle farm, which alternates moments of peacefulness with others of extreme agitation, for both humans and animals. Probably, many people are not always aware of the unpleasant part of the daily routine that, in many cases, is characterized by hard work and serious risk of accidents for humans and animals. For example, many bulls face pain and discomfort due to castration without anesthesia (Fraser, 1972). Added to this unpleasant situation is the stress caused by the handling procedures that precede the surgery, such as lassoing, restraint, knocking down, and tying. Such practices are very common, and they have been justified by economic arguments. Time and again, the same arguments are used to justify not making any changes unless it results in an increase in the farm income.

For those that succumb to this argument, it is important to point out that there is no need for big investments of capital or changes in the market conditions to improve the interactions among humans and cattle during handling. All that is necessary is to learn more about cattle, and to adapt the handling procedures to their nature. Even in more favorable conditions, for example in farms where technological resources and trained people are available, many things can be changed, particularly in the daily handling routines. The first step is to establish ethical principles, which assure the supply of healthy and best quality products, obtained through techniques that guarantee an improvement in the welfare of animals.

The development of best practices of handling

During handling procedures, priority should be given to good interactions between humans and cattle. The aim of applying the best practices is to avoid undermining the welfare of animals during the handling procedures, by minimizing the risk of stress and injury. As a consequence this can be expected to reduce economic losses resulting from cattle morbidity and death.

Handling newborn beef calves

To obtain the highest efficiency in beef cattle production, every cow should ideally give birth to a calf every year, and calves should survive and be in good health.

Obviously, this cannot always be achieved, since in farms around the world some cows do not get pregnant, some calves get sick, and some of them die. Nevertheless, if we want to minimize the rates of calves morbitity and mortality (and the resultant economic losses), we must identify the causes of these problems in order to find the solutions.

It is expected, for example, that very fat and very thin cows will have the greatest risk of having calving problems. At calving overweight cows usually have weaker contractions than those in normal body conditions and thin cows are often unable to face the energy demands on their bodies during calving. One of the major causes of failure (or delay) in first suckling is problems during calving. As part of risk assessment, it is necessary to identify the critical control points, and to work to prevent or minimize their negative effects.

Sometimes a critical control point is defined by the expression of a specific behavior. For example, cows usually lie down when calving and they stand up just after delivering the calf, resulting in the rupture of the umbilical cord. The identification of this simple behavior is relevant because, as described by Schmidek et al. (2006), the risk of calf death is higher for a cow standing up when delivering than for that lying down (16.1% and 4.2%, respectively). However, this information alone is not enough to solve the problem. We have to be able to characterize the situations that increase the probability that a cow will stand up when calving and to take action to avoid them. The authors reported that many factors were important in determining a cow's posture during calving when delivering her calf, among them being the presence of potential predators (e.g. dogs and vultures), the body condition of the cow, and lack of calving experience.

To predict problems with parturient cows and neonates (e.g. calving problems, low maternal ability, low vigor of the calf, failures and delay in the first suckling), it is necessary to carry out routine inspections in the maternity area. This should be done even before the beginning of the calving season, so that the conditions of the ground and the fences can be checked to minimize the risks for the newborn calves. For example, it is common to use small paddocks as calving areas, but this increases the probability that a cow will give birth close to a fence and this can result in a calf falling down on the other side of the fence when trying to stand up, which can result in failure or delay in the first suckling. The inspections should be done at least twice a day (early in the morning and later in the afternoon), all through the calving season.

Under normal conditions, calving takes between 30 minutes and 4 hours. The expulsion of the placenta usually happens from 4 to 5 hours after the birth, and if it takes more than 12 hours, is an indication of placenta retention (Gomes et al., 1998). In general the cow ingests the placenta just after delivering it, and this seems to be important for the definition of its status of welfare (Kristal, 1980).

IMPROVING THE WELFARE OF CATTLE: PRACTICAL EXPERIENCE IN BRAZIL

The ideal is that the calf succeeds in suckling for the first time by the time it is 3 hours old (Schmidek et al., 2006). However, there are some situations that can delay the first suckling. For example, a cow calving for the first time is more stressed by the calving process and this could delay the first suckling or even cause it to abandon its offspring due to interference from a dominant cow (Lidfors, 1994, Toledo, 2005). To avoid this, we recommend that cows calving for the first time are kept in a separate pasture from older cows (Paranhos da Costa et al., 2006a).

Handling beef cattle for vaccination

Vaccination is a common strategy for assuring better health for animals, by minimizing risk of diseases and consequent economic losses. However, it is itself an aversive procedure that often leads to stress in cattle, since they usually have to be handled as part of the process.

Most of the Brazilian beef farms carry out the vaccination in linear chutes, working with groups of animals (usually up to 10) without individual restraint. This handling strategy (named here conventional handling = CH), results in an increase in frequency of cattle jumping, lying down, and trampling on other animals. This makes the cattle difficult to control and leads to an increase in the risk of accidents and welfare problems for both humans and animals (Chiquitelli Neto et al., 2002). Therefore, vaccination should be done in a better way, and we need to look for a reduction in the negative effects of conventional handling.

We developed a new strategy of handling cattle during vaccination, which we called rational handling (RH). This involves vaccinating the animals one by one in the squeeze chute and restraining each animal separately. The aim of this procedure is to increase the control of cattle behavior, and it was hypothesized that its adoption would provide direct economical benefits, decreasing vaccine loss, equipment damages (broken syringes and needles), besides the benefits for the welfare of humans and cattle.

In order to test this hypothesis, we set up an experiment to compare the two handling strategies, CH and RH. The trial took place in a commercial beef farm (Fazenda São Marcelo), where the same team of stock people that usually vaccinated the animals in the liner chute (CH) was trained to vaccinate them in the squeeze chute, applying the RH procedures, and when they were well prepared, the experiment, comparing the two vaccination processes, started.

The results were positive (Chiquitelli Neto et al., 2002). The adoption of RH significantly decreased the frequencies of animals jumping, lying down and trampling on other animals (Fig. 14.1), as well as the frequencies of bleeding, broken equipment, failed injections, and vaccine lost (Fig. 14.2). There was hardly any

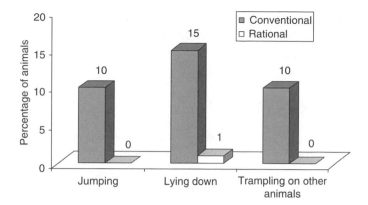

Fig. 14.1   Effects of the adoption of rational handling during vaccination on the percentage of cattle jumping, lying down, and trampling on other animals (adapted from Chiquitelli Neto et al., 2002).

difference in the time spent doing the vaccination (10.2 and 9.8 seconds per animal, for CH and RH, respectively).

Based on these results, the efficiency of RH was tested in six other commercial beef farms to validate the procedures, and to give a detailed description of all behavior. On the basis of all of these results, a guide to best practices of cattle handling was formulated (Paranhos da Costa et al., 2006b). This has now been adopted in many Brazilian beef cattle farms. The farmers that adopted the

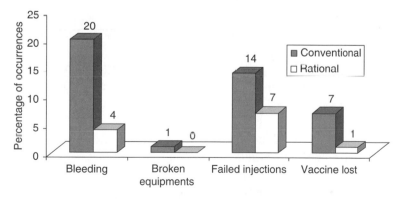

Fig. 14.2   Effects of the adoption of rational handling during vaccination on the occurrences of bleeding, broken equipment, failed injections, and vaccine lost (adapted from Chiquitelli Neto et al., 2002).

IMPROVING THE WELFARE OF CATTLE: PRACTICAL EXPERIENCE IN BRAZIL

RH procedures during vaccination confirmed that it improved the welfare of both humans and animals through the reduction in accidents and aggressive handling.

Handling suckling dairy calves

The high level of morbidity and mortality of dairy calves represents important economic and ethical issues. It is estimated that 75% of deaths in dairy calves occur before 28 days of age (Machado Neto et al., 2004; Coelho et al., 2005), suggesting that calf survival depends directly on the conditions that they face during the early days of life. Even those dairy farms that adopt high technology for milk production face the challenge of high calf mortality rates, which are responsible for significant financial losses.

This was the case for Germânia Dairy Farm, which had high rates of morbidity and mortality of calves, probably (according to our first calves welfare assessment) because the calves under conventional handling procedures (the usual handling adopted by the farm) faced space restriction, inappropriate bedding, and lack of opportunity for social interactions; besides that, the interaction with humans was almost nonexistent.

The strategy for reducing the problems was to adopt the best practices of handling as a routine on the farm. Besides increasing the space available for each animal and changing the bedding material, the main change was to improve human–calf interactions. It was hypothesized that these changes would reduce the calves' morbidity and mortality rates.

We compared the two handling strategies: conventional (CH) and rational handling (RH) (Silva et al., 2006). In the first trial, we used 24 calves (12 per treatment) to compare the effects of CH and RH on calves performance and behavior. The amount and origin of milk and solid food were kept the same for both treatments.

The main differences between the treatments were: RH calves had the opportunity for social contact (the calves were together for at least 2 hours per day in a small paddock); RH calves had better interactions with humans (mainly during suckling, when they received tactile stimulated); and RH calves drank milk from an artificial nipple.

After the first results of the experiment, the farmer decided to adopt the RH. Doing this gave us the opportunity for a long term comparison, using the farm data files where deaths, diseases, and treatments were regularly recorded. Using this information we calculated the frequencies of antibiotics treatments (FAT) and death (FD) over 2 years, then FAT and FD were compared monthly considering 1 year before and after the adoption of RH.

As expected, RH resulted in a clear reduction in calf mortality rate from $6.67 \pm 3.85\%$ to $2.25 \pm 2.21\%$ for CH and RH, respectively, and in the frequency of antibiotics' use with 36.2 ± 14.1 and 8.51 ± 14.78 treatments per month for CH and RH, respectively. These results together suggested that the RH calves were more resistant than CH calves, indicating an improvement in the welfare of the former. Additionally, the adoption of RH resulted in economic gain, due to the reduction of morbidity and mortality rates, which have important roles in the definition of the farm profitability, as reported previously by Bakheit & Green (1981), Hall (1987), Bellinzoni et al. (1989), and Machado Neto et al. (2004).

The welfare of cattle during pre-slaughter handling

Farming activities make a major contribution to the Brazilian economy. Rates of growth in the agricultural sector have been above the average, even during periods of low economic growth. Beef production has a prominent role in this scenario: in order to supply the market 45.4 million cattle are transported and slaughtered every year.

There is much evidence that cattle face welfare problems during pre-slaughter handling (Grandin, 1993; Barbalho et al., 2006), resulting mainly from stress and suffering during loading, transport, and unloading. Besides cattle welfare problems, inappropriate handling also causes a high frequency of bruising and alterations in beef pH, which reduces meat quality (Voisinet et al., 1997).

In order to identify the critical control points during pre-slaughter handling procedures in Brazil, and its relationships with reduced cattle welfare and the occurrence of bruises in the carcass, we set up an observational study (Paranhos da Costa et al., 1998). All situations that would result in these problems were recorded from loading until stunning, and five main problems were identified: (1) direct aggression, (2) inappropriate handling, (3) inappropriate facilities, (4) roads and trucks in bad conditions, (5) cattle being highly reactive. Even under good transport conditions and over short distances, transported cattle showed signs of stress and bruising of varying intensity.

These results were later confirmed by Tseimazides et al. (2005) and Barbalho et al. (2006). In both these studies, the effects of training programs for cattle welfare were evaluated, one for truck drivers and one for lairage workers. Significant reductions in aggressive actions by humans (e.g. use of electric goads) and in stress-related cattle behavior (e.g. falls and vocalizations) were observed. The results indicated that training was more effective when improvements in facilities and labor conditions were also provided.

IMPROVING THE WELFARE OF CATTLE: PRACTICAL EXPERIENCE IN BRAZIL

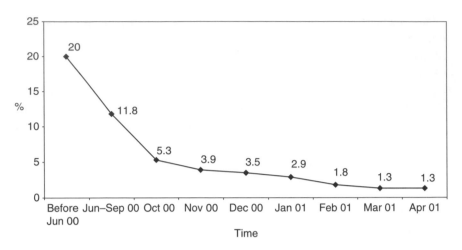

Fig. 14.3 Effects of the adoption of rational handling, showing a reduction in the percentage of bruised carcasses after successive training sessions (adapted from Paranhos of Costa, 2004).

Conclusions

We conclude that to ensure the success of programs to improve cattle welfare, it is necessary to identify the critical control points and to establish a program to improve facilities and handling procedures. Such an approach has been developed by our research group, Group ETCO (Ethology and Animal Ecology Research Group), in partnership with farmers and private companies that expect the studies to solve their practical problems and also offer structural and financial support to carry on the research. The knowledge is transferred to the production sector through courses, workshops and consultancy, communicating directly with farmers, stock people, drivers, and lairage workers.

   The results are clear cut. Improved labor conditions, better animal welfare and lower economic losses are seen following the adoption of rational handling. For example, there was a clear reduction in carcass bruising after the adoption of rational handling during pre-slaughter handling (Fig. 14.3), which also represents less pain and suffering for the animals. The way ahead must be to promote both human and animal welfare together with, as stressed elsewhere in this book, protecting the environment.

BRINGING ABOUT CHANGE

15 Organic farming and animal welfare

After graduating from Harper Adams Agricultural College in 1986, **Helen Browning** *took over the tenancy of Eastbrook Farm from her father and began the conversion to organic farming. She started Eastbrook Farms Organic Meat in the late 1980s and opened a butchers shop in the nearby village of Shrivenham. Since then the business has grown and her organic meat is now available to customers through a nationwide home delivery service and via supermarkets. She was appointed as Food & Farming Director at the Soil Association in 2004 and also chairs the Food Ethics Council and the Animal Health and Welfare Implementation Group for England. She is a Meat and Livestock Commissioner and was a member of the UK Government's Policy Commission on the Future of Farming and Food ("the Curry Commission").*

As the realities of climate change and fossil fuel depletion finally burn their way into the public's consciousness, there is a serious danger to some of the progress made on animal welfare. Already there is discussion of the desirability of shortening the lives of beef cattle, with the inevitable intensification of production systems that that would bring, to reduce methane emissions and improve lifetime feed conversion. Free-range now spells inefficient, as pigs and poultry roaming and romping burn up calories as they enjoy something approaching a worthwhile existence. Inevitably, the GM advocates begin to bludgeon us with promises of animal adaptations that will improve their carbon/greenhouse gas footprint. Who are we to object in the face of this global emergency?

The ironies of these new attacks on a gentler, more natural approach to animal husbandry are legion. The whole point of developing an aspirational system of farming that we now term "organic" was the reconciling of the myriad strands of need and care of planet, people, and animals into a practical approach that, whilst not perfect, has perfection as its aim. The fundamental principles, of minimizing resource use, of recycling through the soil thus building organic (i.e. carbon-based) matter therein, of positive health – a state of immune competence and vitality – all these, and many others, are the foundations on which all of agriculture, indeed, all of society, should be based. Instead, arrogance and wastefulness and dominion have been our calling cards, and now we are in a right old pickle. And instead of learning that we must work within the principles of the natural world if we want a long term future on this earth, we push harder at the wheel that breaks the butterfly.

The organic approach to animal well-being is to integrate livestock into farming systems so that they both benefit the land and benefit from it. The behavioral needs of livestock are particularly emphasized and restrictions on the use of medications aim to ensure that systems must be built around good husbandry – plentiful space, a diet compatible with the animal's physiology, the ability to exercise freely, lower production levels, later weaning, constraints on mutilations are a few examples. The differences are most marked for pigs and poultry, the animals which have suffered most from increasing intensification. Organic pigs, for instance, must be free-range throughout their life (unless weather conditions are extreme); they move across land rapidly preventing parasite and disease build up and minimizing the risk of nutrient leaching from the land; they cannot be weaned earlier than 6 weeks, ideally 8; teeth and tails are never clipped – they don't need to be.

The organic movement has achieved much in developing farming systems with a high welfare potential, and in creating a credible consumer proposition that has gained considerable momentum in the marketplace. This proposition is about a whole range of benefits, and although a key issue for many purchasers of livestock products is the welfare message, the "bundling" of virtues – including biodiversity, carbon footprint, fairer trade, minimization of residue risk, health and food quality – maximizes appeal and relevance to a wide audience. All well and good, but the organic movement faces a number of internal and external challenges that must be acknowledged and addressed. Some of these challenges are discussed in more detail below.

Ongoing standards development

The standards setting process has always had to acknowledge real world constraints (such as the availability of organically grown protein crops, or of appropriate broiler genetics) as well as knowledge gaps and occasional trade offs between the realization of the wide ranging objectives of organic principles. The rate of progress in standards development has been slowed by the introduction of the EU regulation, which, whilst helpfully ensuring a baseline across Europe and thus facilitating trade, means that agreement of improvements is long winded and subject to both commercial pressures and member state priorities. The Soil Association, a membership charity whose trading subsidiary certifies the majority of UK organic food, has fought hard to maintain its right to set standards at a higher level than the EU baseline. The need for this is especially necessary in controversial, difficult areas like pigs and poultry. Even without the legislative challenge, moving ahead of the game whilst keeping producers, retailers and the public on board requires evidence of benefit, the ability to demonstrate new approaches and support change, and to communicate positively with the public – ideally without completely antagonizing the rest of the "industry."

Ensuring that system potential is achieved in practice

Standards set the scene, define the system, prevent systematic welfare abuse (such as confinement, routine drug use, inadequate housing, unnecessary mutilations) and embed good practices (like late weaning and appropriate diet). What standards cannot guarantee is good stockmanship. The change of mind set, as well as new practical knowledge, that the stockperson needs to be able to take on board presents a real challenge and requires considerable support in many cases.

The move towards the measurement of welfare outcomes, which the Soil Association, working with Bristol University, is now putting in place as part of its inspection process, is key to ensure that welfare benefits and outstanding problems are clearly understood and, where necessary, tackled. Welfare outcomes cannot substitute for "input" standards, and need to be seen in the context of the system and its benefits to the autonomy and freedom of the animal. But the monitoring of welfare outcomes will provide the means to ensure improvements in management and to highlight both standards development and advisory needs.

Commercial pressures

Inevitably, the strength of the organic market has attracted many whose initial motivations may be more commercial and less in tune with organic values. While big is by no means necessarily bad, and accepting that scaling up and improving the "efficiency" of organic systems is likely to be important to minimize the price to the consumer, inevitably tensions will continue to arise. These can only really be resolved by a clear understanding of the public interest, by honest communication of the attributes and costs of different farming systems. Maintaining public trust and confidence is paramount if the opportunity to use the market mechanism to substantially improve animal welfare is not to be jeopardized.

ORGANIC FARMING AND ANIMAL WELFARE

Localizing our food system

Although the climate change/mitigation agenda has been shown to hold some threats to progress achieved to date for farm animal welfare, there are considerable opportunities also. The need to reduce the impact of our food systems beyond the farm gate brings the challenge of reducing emissions in transport, processing, packaging, and distribution. In these areas, organic systems are delivering as little as "conventional." To achieve substantial improvements in this regard, local infrastructure for the slaughter, processing and distribution of food, including meat, will be essential. This could lead to gains regarding shorter distances to slaughter, and a move towards more varied farming enterprises, often with smaller herds and flocks, which are better suited to feeding people more directly and locally. As some welfare problems may increase, in larger enterprises, this could bring benefits. An example of this is the growth in interest in micro-dairies, typically comprising 10–30 cows, where the diseconomies of scale on the farm are compensated for by the additional value added in processing and selling direct.

The prospects for substantial improvements for animal welfare, then, hang in a balance between the threat of societal upheavals and changes which may lead to both policy pressures and a more "selfish" attitude amongst consumers, and a vision, which the organic movement must continue to play its strong role in developing and maintaining, of a future where humanity recognizes that its needs and those of the planet, including its other inhabitants, are inextricably linked. Only by this reconciliation can we achieve a sustainable future for humankind.

BRINGING ABOUT CHANGE

16 Enlightened Agriculture and the new agrarianism

Colin Tudge read zoology at Cambridge in the early 1960s and ever since has earned his living by writing, mostly on aspects of biology but largely, too, on food and agriculture. He has now published around fifteen books and in the latest, Feeding People is Easy, *he shows that the present disasters in the world are caused entirely by bad policy, and discusses the things we need to do to put things right. He is a director of the Food Ethics Council and a trustee of the Farm Animal Welfare Trust.*

It should be fairly easy to feed everyone who is ever liable to be born into this world well, and forever; and by "well," I don't just mean "adequately" – enough calories to get through the day – I mean to the highest standards both of nutrition and of gastronomy. We do not need to live austerely, on beans and cereals. We – collective humanity – merely need to take food seriously. The future, in short, belongs not to the ascetic, but to the gourmet.

Can it really be so? If it is, then why do governments, industrialists, bankers, and intellectuals and experts of all kinds – economists, accountants, scientists, technologists, lawyers, and business managers – queue up to tell us the precise opposite? Can it be that they have a vested interest in the status quo? Can it be that they have got it wrong?

In truth, my case is easy to demonstrate. The present world population stands at 6.4 billion, of whom 1 billion are undernourished and 1 billion are dying before their time of various surfeits (of fat, sugar, salt, and so on). By 2050, the United Nations tells us, there will be 9 billion of us, or thereabouts. But the curve of growth is flattening out, and by 2050 numbers should stabilize. In the fullness of time, provided we do not pursue the current, somewhat obscene fantasy to prolong individual lives indefinitely, total numbers should go down. So the task is to provide for 9 billion people by 2050, and to go on doing so forever.

Can this be done? Well – the world already produces enough food for 9 billion people so of course it can be done. Global warming is a huge threat and perhaps it will get the better of us but it seems most likely that if we cannot in the future grow food in the places where we grow it now, then we ought to be able to grow it somewhere else. Some places will get drier but some will get wetter and although we may lose much of the coastal plain, which contains much of the best farmland, we should certainly gain elsewhere. We should of course be trying very hard to minimize global warming and we might in the end fail but there is no obvious reason simply to give up. With the climate as it is today, even with farming that in large part is truly dreadful (and getting worse), we provide enough food to feed as many people as there are ever liable to be at any one time. So with halfway sensible farming we should be able to do the job even though the climate is so precarious.

All we need, in fact, is agriculture that is actually designed to feed people. I call this "Enlightened Agriculture": a big and fancy term for doing the thing that is obvious, in very obvious ways.

The secret – it really is so obvious I feel mildly ashamed to write it – is to ask what human beings really need and want; then to ask what the world could produce, if it was farmed sensibly; and then see if need and output can be made to match up. In truth, they seem to match up beautifully.

How, first of all, do we produce enough? What does "halfway sensible" farming entail?

We begin with the idea of biological reality. Politicians, industrialists, bankers, economists, and accountants are forever telling people who want to do things differently to be "realistic." But to them, "realistic" means doing the things that accord with the status quo; with present-day institutions, and with present-day economic models. The economic model that now prevails worldwide is that of the global "free" market which is driven by the perceived needs first to generate as much wealth as possible in the shortest possible time, and second to compete – so it isn't enough to be rich; you have to contrive to be richer than everybody else, and preferably put the rest out of business. The overall statistics looked at honestly show that the ranks of the world's poor are growing by the day but the present way of organizing our affairs is, we are assured, the best that can be done. The rules of the status quo now have the power of global law, enshrined in particular in the World Trade Organization, and so it is "unrealistic" to propose anything different.

But in truth, the only "real" reality is the physical reality of Earth itself, the place where we live, and the biological realities of the creatures it contains, including ourselves. Money, by contrast, is an abstraction. That is how money was first conceived, many thousands of years ago, and that is how it remains; not reality itself, but a symbol, a token of reality. The politicians, industrialists, bankers, and economists who now run the world and seek to gear all our efforts to the generation of cash have mistaken the token for the things it betokens. For them, the token has become the reality. Most of the world's trade these days is not in goods but in money. Exactly one half of Britain's national wealth is currently generated by the exchange of money. The world is failing, obviously and horribly before our eyes, for the very simple reason that it is rooted in an abstraction – that of money; which in the eyes of the people who run the world takes precedence over the world itself.

Biological reality, by contrast, acknowledges that the world and all its components are finite – land, fresh water, energy; that the climate will do what it will, strive though we might try to shape it to our convenience; that landscapes too, in the end, are as they are – and of course extremely various; and plants and animals have physiological limits which we can reasonably extend to some extent for our own benefit – but if we take too many liberties with them then it becomes both dangerous and cruel.

The basic task is to produce enough calories, protein, and micronutrients of many kinds to feed the human race without wrecking landscapes or wasting all the fresh water and fuel, and without undue risk and cruelty. The reasoning that follows from this is simple and obvious. Protein and calories are produced most efficiently by growing staple crops on a field scale: cereals, pulses, tubers such as the potato, and various nuts including the coconut. This is arable farming, from a Latin word meaning "ear" as in ear of wheat. The most favored land of all is reserved for horticulture, in which plants of as many kinds as we find desirable are grown, in effect, plant by plant: "horticulture" is of course from the Latin *hortus*, meaning garden. Horticulture in the main provides micronutrients. Livestock is fitted in where it can be fitted. This is pastoral farming, from *pastor* meaning care. The ruminants, mainly cattle and sheep, feed on grass or – especially in the tropics – on the leaves and branches of trees (which is "browse"). They are kept mainly on land that is too steep or hot or cold or wet or dry for arable farming; and they also graze on short term "leys" in arable fields, in periods of rest between crops of staple, providing some fertility. Pigs and poultry feed basically on leftovers and crop surpluses; and since the aim in any one year must be to produce a slight surplus (because crops vary and we need a safety margin) there should usually be a surplus, and there should always be a store to ensure continuity.

The end result of such biologically realistic farming is to produce lots of plants, not much meat, and huge – maximal – variety. In overall structure, this kind of farming reflects nature, which also produces a lot of plants with relatively few animals; and so in principle such farming is as stable and sustainable as nature itself. This is also, of course, the basic structure of traditional farms – which of course are enormously various worldwide, and from century to century, but basically are structured as described here: arable and horticulture taking pride of place, with animals fitted in where they can. "Enlightened Agriculture," in short, is structurally the same as traditional agriculture as practiced for at least the past 3000 years in all but the most special environments. It is common sense, and it would raise no eyebrows at all among people who know anything about farming.

ENLIGHTENED AGRICULTURE AND THE NEW AGRARIANISM

Now we come to a series of wondrous serendipities – again blindingly obvious; but revelatory in a world in which the obvious has been obscured beneath layers and layers of vested interest and rarefied economic theory. Enlightened Agriculture – agriculture designed to produce as much food possible within the bounds of biological reality – produces lots of plants, not much meat, and maximum variety. These nine words – "lots of plants, not much meat, and maximum variety" – summarize all the nutritional theory of the past 30 years that is worth remembering: all those thousands of pages of learned reports, and health guides, and what you will. In short: the products of Enlightened Agriculture, designed to respect the biological realities of landscape, crops, and animals, perfectly match the nutritional needs of human beings. Actually it could hardly be otherwise. Human beings are supreme omnivores. We are adapted to eat what nature provides, in the proportions that nature provides it. Enlightened Agriculture, which in essence is traditional agriculture, is in essence an abstraction of nature.

But although "lots of plants, not much meat, and maximum variety" is what people need, is it what they actually want? Don't they want meat, meat, and more meat? Isn't it the duty of all enlightened and altruistic food industrialists to produce an indefinite quantity of hamburgers and fried chicken? Isn't this what "the consumer demands"?

This is arrogance or ignorance or both writ very large indeed. In truth, "lots of plants, not much meat, and maximum variety" is the basis of all the world's greatest cuisines: Provencale, Italian, Indian, Chinese, Turkish, Lebanese, where you will. These cuisines are democracy in action: they are by the people, of the people, and for the people. They have never been significantly improved upon, and never can be – as all the world's greatest chefs acknowledge; all the greatest are in search of traditional flavours and dishes. The products of the modern food industry by contrast are like – well: Kiddicraft against the Taj Mahal. But for this foulness, this coarseness, the glorious tradition that could serve us all so well is being systematically trashed.

In short, just to over-egg the point, the kind of farming that could ensure a steady food supply forever and would feed us to the highest standards of nutrition, would also raise us – all of us – to the highest standards of gastronomy.

Enlightened Agriculture is rooted in social realities, too. For what are 6.4 billion people (or 9 billion people) supposed to do with their lives? Two hundred years ago, in Britain, the answer was obvious. There was coal galore, and an entire Empire to draw upon, and roaring new technologies for manufacture; and men, women, and children were needed in superabundance to work in the factories. This lasted until well into the twentieth century. But that particular party is over. The rest of the world is industrializing too, of course, but there aren't enough resources in the world for them all to repeat the industrial transformation seen in Britain and then the US. As Gandhi asked disingenuously half a century ago, "Who will be the Third

World's Empire?" Besides, factories on the whole were not nice, and most of the jobs in them have been taken by ever more intelligent machines. So what is the majority of humanity actually to do?

Well, half the world's population – about 3.2 billion – now live in cities; and about one third of them (an estimated 1 billion) live in slums. Some say that the slum economy is "vibrant." People make wondrously ingenious toys and shopping baskets and goodness knows what out of old polythene bags and tin cans and tyres. They dress each others' hair. There is prize-fighting, mugging, begging, drug-running, and prostitution, and the young and fit can always find a war to fight in. It's a full life, in short. The luckier ones (if such they be) get to clean hotels – in Brasilia they are bussed in and out each day, before dawn, lest the tourists and diplomats should catch sight of them. They may drive taxis – 80 hours a week in Delhi for £8 a month. If they work very hard, and pass their exams, they get to work in a call center. It is much the same in the US, where millions of poor people hold three jobs (cleaning, fast-food, cleaning again) and just about manage to pay the rent. That is the new economy for a very large proportion of the human race, and as things are it cannot get any better. Apart from cleaning hotels, and driving taxis, and fighting and whoring, and serving burgers, there is nothing that the modern economy requires of them – except that they should continue to consume the things that the modern industrialists find it convenient to produce.

Traditionally, though, most of the world's people have been engaged in agriculture. Agriculture is still the world's biggest employer by far. Worldwide, around 40% of the workforce are on the land. In the Third World as a whole it is 60%. In India, specifically, 60% of the workforce with their families live on the land – a total rural population of 600 million. In Angola and Uganda it is 80%; in Rwanda, 90%.

Modern industrialized agriculture doesn't need people. In Britain and the US, only about 1% of the total workforce are full time on the land. Thomas Jefferson conceived the US as a nation of small farmers but now there are more people in jail than full time on the land. This extreme reduction of labor is considered highly efficient, because it is cheap; in traditional systems, labor is the most expensive input. A Lincolnshire farmer told me recently that he employs one man on 1000 hectares, but would prefer to employ one man on 2000 hectares. This is fake of course: the US has always supplemented its agricultural workforce, first with slaves and then with immigrant labor – and continues to rely on immigrant labor. So too, increasingly, does Britain. The agricultural communities of George Eliot and Thomas Hardy are replaced by east Europeans, bussed around the country like Brasilia's hotel cleaners. This is progress.

Because this is progress, this is the course that is advocated by the G8 governments, and by industrialists and bankers, as the model for the whole world to follow. But if India followed Britain's farming model, then 600 million people

ENLIGHTENED AGRICULTURE AND THE NEW AGRARIANISM

would be dispossessed – a population far larger than the whole of the present EU even in its expanded form. About 60,000 of them (1 in 10,000) might find a job in India's new and excellent IT industry – if they had degrees, which of course they do not. The rest? Well – there are hotels to clean, taxis to drive, and the slums. I have been to the slums. Vibrant they aren't.

In truth, the only realistic occupation for most of the world's people in the future is to work on the land, as most people have done this past 3000 years. And here, Enlightened Agriculture provides us with yet another serendipity. Farms that are designed to be maximally efficient in biological terms (as opposed to cash-efficient), and to operate sustainably within the caprices of local landscape and climate, must be mixed. They require judicious combinations of livestock and crops, with various species of each, integrating pastoral, arable, and horticulture (and spreading too, though not too profligately, into the wild environment and encouraging agroforestry). I have seen such farms recently in China – paddy fields to the horizon, but with intensive horticulture on the islands of raised grounds, and ducks feeding on snails and parasites among the stalks of rice and fertilizing as they go, with pigs and chickens in the village, and the occasional water buffalo to help out. In Britain traditional farms combine sheep, cattle, wheat, barley, and a cottage garden – but the principle is the same.

Such farming, though – designed primarily to produce good food – needs high levels of husbandry: a lot of people, who know what they are doing. It needs to be labor intensive.

In short, the only kind of economies that truly can serve the needs of humanity in the future must be rooted in agrarianism. This, of course, is the precise opposite of what today's governments and industrialists and their attendant experts and intellectuals have been telling us. To them, agrarianism is "the past," and the anathema. "We cannot turn the clock back!" is the standard cry. Progress means no more farms of the kind that most of us still recognize as farms.

To be sure, we should not simply turn the clock back. But the kind of agrarianism that I am envisaging as the necessary grounding of Enlightened Agriculture is not a simple reconstruction of the past. I prefer to speak – yet another slogan – of "the New Agrarianism." We all know that agrarian living can be extremely hard – physically hard, and mentally debilitating because rural people tend to be cut off from the rest of humanity. Worldwide, rural people have been looked down upon, as yokels and hicks, and in general have low status. To be sure, many of the people who do the looking down are hicks themselves, albeit urban hicks, but that is not the point. Agrarian life is intrinsically difficult in many ways – but the task for humanity is to make it agreeable. The main ways of doing this, as ever, are social: a change of attitude. Good farmers should be admired, just as doctors and teachers were tradition-ally admired. An economic shift is needed too, of course – people in general need to pay more for food, to ensure that those who produce it are not done down.

But also, the new agrarianism needs science and the high technologies that come out of science. "Appropriate technologies" are not necessarily "low" technologies. To be sure, a great deal could be achieved in many parts of the world with better harnesses for draft animals, and smoother axles on carts – but the highest of high tech has many roles too. Indeed, it is often the poorest people who can benefit most from the highest tech. The world needs far more subtle methods of pest control – the blast of industrial chemistry just won't do; and this is a huge challenge. More heat-resistant vaccines for tropical cattle are always needed. Perhaps above all we now have the internet. People who live in remote places need no longer feel cut off. You can enjoy the silence of the Scottish Highlands or the Brazilian Cerrado if that is what suits you – or, in principle, while you tend your sheep or cultivate the semi-wild fruit trees, you can tune in to Oxford and continue your PhD. We could, with modern technology, have the best of both worlds.

Ivan Illich in the 1960s coined the right expression: not simply "appropriate technology" but "tools for conviviality"; technologies that make life more agreeable for human beings in general – and, most importantly of all, technologies that increase individual autonomy, making each of us less dependent on governments and industrial companies, and especially on industrial companies that may be based in another continent. Agriculture always has been, is now, and always should be a craft industry: dependent on the genius and the efforts of individuals. The role of science should be to abet that craft: "Science-assisted craft" is yet another of my little slogans (to lie alongside Illich's "tools for conviviality"). At present, agricultural science is not designed or intended to assist the crafts of farming. It is intended and designed to sweep them away; to replace them with the formulae, the algorithms of industrial farming: farming out of a bag; "farming by numbers"; husbandry that is as simple as possible – monocultures rather than mixed cultures – and employing as few people as possible. Science in short has lost its way. Humanity needs to reclaim it; to rescue it; to turn it again to human purpose.

The world needs a debate, of course – none more vital: What is the ideal proportion of people working on the land? We can all agree that Rwanda's 90% is too many – the Rwandans certainly agree with this: with only 10% in the cities, there is no one for farmers to sell their produce too, so they are doomed to perpetual subsistence (the idea that they could realistically sell their produce profitably on the world market is just another of the modern economists' sick little jokes). On the other hand, the 1% in Britain and the US is obviously too few. Leaving aside the social unpleasantness – the few farmers who are left are lonely, overworked, and increasingly desperate, and the immigrant laborers are not exactly happy either – it is extremely risky. The smouldering epidemic of BSE in Britain (the human infections have yet to run their course) and the foulness of the recent foot and mouth epidemic resulted entirely from cut-price husbandry. They were not acts of God. If we go on as we are, we are doomed to lurch from epidemic to

ENLIGHTENED AGRICULTURE AND THE NEW AGRARIANISM

epidemic. But in the modern economy the agriculture of Britain and the US are perceived to be the endpoint to which all the world should aspire. Adam Smith, in the eighteenth century, pondered this issue – how many people *should* work on the land? We should not have stopped pondering it. But apparently we have. Just to set debate in train, I suggest that no country should ideally have more than 50% of its population on the land – but none should have fewer than 20%. This would mean that present-day India is just about right, while Britain and the US have gone out on a disastrous limb.

Why has the world got itself into such a mess? Why do the politicians, industrialists, bankers, and all their attendant experts and intellectuals fail to see what is so obvious? Why do they continue to lead us into such an obvious cul-de-sac?

Of course the answers are many and intricate – well worth exploring, but not here. The proximal answer is that the world has locked itself in to a very peculiar economy. The fault does not lie with capitalism per se. Thomas Jefferson and the other great founders of the modern United States perceived the US as a capitalist state, rooted in free enterprise and free trade. But they envisaged free trade as envisaged by Adam Smith: a very large number of individual traders doing their thing, out of which an "invisible hand" miraculously creates equity and justice – an early exercise in complexity theory. Smith saw the corporates – huge amalgamations of ever increasing power conceived as engines of wealth – as the enemies of free trade. He knew all about corporates – he had the model of Britain's East India Company before him. Jefferson and his colleagues knew all about corporates too – the excesses of the East India Company triggered the Boston Tea Party of 1773 which led in short order to the Declaration of Independence of 1776. Jefferson and James Madison in the early nineteenth century placed specific restrictions on the growth of corporates. Those restrictions form part of the US Constitution. But although Jefferson's and Madison's restraints remain in place, theoretically, the corporates slipped through the net; and now the entire world economy is dominated by corporates. Governments like those of modern Britain and the US can properly be seen as extensions of the corporate boardroom.

The exchanges of corporates and superpowers, half mortal combat and half cabal, cannot be expected to produce a world with the delicacy of touch required to attend to the well-being of humanity. Why should it? In fact, the requirements of the modern corporate–superpower economy seem designed expressly to kill off the kind of farming that we need, and the social structures that go with it. They demand maximum productivity – hugely dangerous and nonsensical if we acknowledge the biological realities of a finite and delicate world, and the need for sustainability. They demand maximum "value-adding" – and this above all means maximizing the output of livestock. So it is that meat no longer helps us to grow more and better food – it competes with us for food. Our livestock currently consumes half the world's wheat, 80% of the maize, and well over 90% of the soya – all staples that could and should be creating great cuisine. By 2050, on current trends, the world's livestock will consume the equivalent of 4 billion people – roughly equal to the total human population in the 1970s, when the United Nations held its first international conference to discuss the world's food crisis. Above all, the perceived need to maximize profit requires producers of all kinds to minimize costs. In farming this means simplified, monocultural agriculture in place of the intricacy that imitates nature; it means racing animals from birth or hatching to the abattoir in the shortest possible time – just a few weeks in the case of broiler chickens; but above all it means the reduction of labor. With that goes risk, and the horrors of unemployment, the royal road to poverty, the thing with which we are officially supposed to be at "war": perhaps the sickest of all the modern economists' sick little jokes.

So what's to be done? How can we bring about the New Agrarianism, constrained by biological reality, based upon Enlightened Agriculture, and driven by Science-assisted Craft? In the long-term we need to dig deep. We need perhaps above all to address morality – we need to give a damn. We need to *care* that the 9 billion people of 2050 should be well fed. It is absurd to suggest that superpowers, bent on their own supremacy, or that corporates focused on their shareholders, have such a target in mind. We need, too, to abandon the present absurd world economy, the global dog-fight in which we are all embroiled. We do not need Marxist or Maoist revolution. We just need to return capitalism to its roots – as envisaged by Smith and Jefferson. The present system is a betrayal.

But if we do not attend to the short term then there can be no long term. What can we do now, to arrest the madness?

In principle, there are three ways to bring about change: Reform; Revolution; and Renaissance.

I would not waste time on Reform. There is no point talking to politicians and industrialists. I have tried. They are not going to change their minds. If they do then they will tell us that they are powerless to act. If we ever had an enlightened Secretary of State for Agriculture then he or she would simply point out that Britain is locked into the WTO and the EU, and that's the end of it. We might legitimately ask at that point what politicians actually think they are *for* – how they justify their salaries – if they cannot do the things that obviously need doing. But such polemic would only waste more time. It is not worth entering such discussions.

Revolution is not advisable. Full-scale Marxist revolution as in Russia and China is not required – proper capitalism is what we need; not centralized economies. It would not be good to burn down Tesco's. People might get hurt, and the incendiarists would get into trouble, and nothing worthwhile would be achieved. In general, revolutions are too difficult to control. There is no knowing how they might turn out.

That leaves Renaissance. This is eminently achievable. There is no need to transform the status quo, or to sweep it aside. Just ignore it. If people who do have sense and ability simply do their own thing, then they can build something new: a new food supply chain, based on Enlightened Agriculture and the New Agrarianism. One way forward would be to establish what I like to call "The Worldwide Food Club." This is (or would be) a cooperative of supplies – farmers, cooks, brewers, bakers. and so on – who are truly committed to the cause of good food; and of consumers, who are prepared to pay for good food, properly produced. The ingredients for such a club are already out there – The Slow Food Movement; Fair Trade; Compassion in World Farming; The Soil Association; countless farmers' markets and small farm cooperatives; and millions and millions of individuals who can see perfectly well what is needed and want only the opportunity to act. So the club could grow rapidly. Without any individual becoming rich –the WWFC should be run as a Trust – it could soon become a significant force. So there is no need to burn down Tesco or the Department of Agriculture. Just create something better, and allow the status quo to wither on the vine.

It is not necessary to persuade people to change their minds. There are enough out there already to make it work. They do not need to form the majority. A critical mass is all that is required to change the world. And here is the final serendipity. The great desideratum, so most people in the world seem to agree these days, is democracy. Humanity's most important material exercise, the thing we absolutely have to get right, is our food supply. Serendipitously, of all human enterprises, the food supply is the one that lends itself most readily to democratic control. We can all get involved as from now. But we have to do it ourselves. Let's go for it. I think it is our only chance.

Coda: Enlightened Agriculture needs thinking through, of course. There are hundreds of details to be worked out – not least because the world's intellectuals have been thinking along quite different lines this past 30 years. So it would be good to establish a "College for Enlightened Agriculture" to underpin the endeavor. Talks are afoot for this.

Colin Tudge discusses all these ideas further in Feeding People is Easy, *published in 2007, and on his website: www.colintudge.com.*

BRINGING ABOUT CHANGE

17 Conclusions

Marian Stamp Dawkins and Roland Bonney

It was never our intention to provide complete answers or to make precise predictions about what the future of animal farming might hold. The future of our planet itself is so much in the balance that being too prescriptive or certain at this stage is unlikely to be particularly helpful. Rather, we wanted to raise questions and pose challenges to existing ideas. By presenting a variety of different views in the different chapters of this book, we wanted to prepare peoples' minds for the new ways of thinking that may be needed as we move towards a changing future of farming.

Consider how very ignorant we are. We don't even know whether there will be any animal farming at all if we look more than a few years into the future. Perhaps we will all subsist on plant protein or food that has been manufactured from entirely synthetic materials. Perhaps. But then again, perhaps not for a long time. It is not even clear where the measurements of "carbon footprints" will take us. Will it take us towards more and more locally grown food so that "food miles" are reduced as much as possible? Or will the energy equations in fact turn out to make the opposite predictions and lead to the concentration of some types of food production in certain areas of the world where it can be grown most efficiently and then shipped, perhaps over long distances, to where it is needed? If food miles are redefined to include not just the distance from farm gate to consumer but all the deliveries that have to be made to the farm (animal feed, water, energy for heating, etc.), will it make more sense to rear chickens in Brazil and then send them to Europe or to send the soya and maize to Europe in order to grow the chicken "locally" in Europe?

And of course, the future of farming will depend not just on its carbon foot-print but on a host of other political and economic factors, let alone other resources such as water. Will terrorism or the threat of disease pandemics put a premium not on growing food where it can be grown most efficiently but on self-sufficiency of food production, with each country attempting to insulate itself from global threats by producing as much as it can for itself, despite the difficulties and inevitable inefficiencies of trying to grow, say, tropical produce at high latitudes?

With such uncertainties about what farming in the future will be like, it is clear that we have to be both very clear about what our priorities are and also very flexible in how we go about achieving them. The message of this book is that animal welfare is one such important priority, although one that is in danger of being sidelined as climate change assumes greater and greater importance in peoples' thinking. But as our various contributors have argued, trying to define sustainable human food production without including animal welfare is unlikely to be successful. Not only are we humans utterly dependent on the health and welfare of animals for our own health and survival (as the threat of bird flu constantly reminds us), but a world of animal abuse is not a world many people want to live in. The quality of human life depends both physically and emotionally on the quality of our relationship with nonhuman animals. Animal welfare therefore has to be as much part of sustainability as environmental protection, food quality, and economic prosperity. To put it another way, "sustainability" has three elements – Economics (affordable food), Environment (a viable planet), and Ethics (what is socially acceptable) – and animal welfare is part of all three.

Making sure that policies on sustainability do include animal welfare, not just as an optional extra but as an essential part of that policy itself, is not going to be easy. It means exerting constant pressure to make sure that animal welfare is, and remains, part of the way we shape the future of farming. In the course of this book, we have seen how an unlikely alliance of big business, farmers, scientists, consumers, and animal welfare organizations is bringing about real change in the way farm animals are treated. Sometimes improving animal welfare is itself enough of a driver to bring about change. At other times, it gets a boost from the increasingly recognized link between what is best for animals and what is best for human health and well-being. Animal welfare thus moves forwards under the impetus not just of itself but as part of the whole sustainability package.

This is only to be welcomed. If animal welfare can ally itself with what is good for human health, what is also good for the environment, and what people believe is an ethical way of living, then it has a real chance of making headway. If happy animals are safer animals as Humphrey (2006) puts it, animal welfare has a chance of being taken seriously even by people who don't take animals particularly seriously.

But of course, we, the contributors to this book, do take animal welfare seriously and do believe that animals matter for themselves. And yet we, too, are going to have to be flexible in what we advocate and even abandon some long-held views about the way animals should be treated. We don't know enough yet about how to farm animals commercially with the highest standards of animal welfare to be able to lay down in precise detail what should go into our new contract with farm animals. It is not enough to respond to the criticisms of intensive farming by simply opening the barn doors, making all animals "free-range," and expecting all welfare problems to instantly disappear. True husbandry that genuinely cares for animals and ensures their health and well-being is more complex than that. Sometimes we already know some of the answers. Existing "best practice" among farmers could be written directly into the new contract. But in other cases, we don't know. We may not know the best diets to give them or we may not yet have the right breeds that are able to thrive outside intensive housing. Or we may not understand their social system or their needs for space well enough to know how to give them the environments they want. Since they cannot tell us in so many words what they want written into their side of the contract, we have a responsibility to find out on their behalf. And that means some degree of humility on our part and an ability to change our minds if the evidence points that way. Insisting that all animals must be "free-range" and outside in a cold European winter may turn out to be less good for welfare than well built indoor systems designed with the animals' own needs and requirements in mind. If we want to achieve the highest standards for animals, we need to apply the same standards for assessing welfare across different systems and not come with pre-set ideas that one system or another must be better for welfare just because we humans think it looks better that way. The terms of the new contract must reflect what really does improve animal health and what really does give them what they want.

We hope you have enjoyed the book and have found it interesting or surprising or informative or thought-provoking or, best of all, all of these. The future of animal farming is more uncertain than perhaps it has ever been and the one thing we can expect is change. In dealing with that change and working for a sustainable future for the planet, all of us who care about animals, and indeed the health of our own species, need to make sure that animal welfare remains firmly at the center of what is meant by sustainable farming. Farming that ignores animal welfare cannot be "sustainable."

References

Anderson, R. (2007) Mid-Course Correction; a story of an eco-epiphany. *Resurgence* May/ June No. 242:36–37.

Appleby, M.C. & Hughes, B.O., eds (1997) *Animal Welfare.* CAB International, Wallingford.

Arey, D.S. (1992) Straw and food as reinforcers for prepartal sows. *Applied Animal Behavior Science* 33:217–226.

Bakheit, H.A. & Green, H.J. (1981) Control of bovine neonatal diarrhea by management techniques. *Veterinary Record* 108:455–458.

Bagley, C.V. (2003) Tail Docking of Dairy Cattle. http://extension.usu.edu/files/agpubs/ taildock.htm. Accessed January 12, 2006.

Barbalho, P.C., Tseimazides, S.P., Naves, G.A., Ciocca, J.R.P., Neves, J.E.G. & Paranhos da Costa, M.J.R. (2006) Programas de treinamento em bem-estar animal: efeitos na insensibilidade de bovinos logo após o atordoamento. In: I Congresso Internacional de Conceitos em Bem-Estar Animal, 2006, Rio de Janeiro. Anais do I Congresso Internacional de Conceitos em Bem-Estar Animal. WSPA, Rio de Janeiro, 2006.

Barnett, J.L. and Hemsworth, P.H. (1990) The validity of physiological and behavioural measures of animal welfare. *Applied Animal Behavior Science* 25:177–187.

Bellinzoni, R.C., Blackhall, J., Baro, N., Auza, N., Mattion, N., Casaro, A., La Torre, J.L. & Scodeller, E.A. (1989) Efficacy of an inactivated oil-adjuvanted rotavirus vaccine in the control of calf diarrhea in beef herds in Argentina. *Vaccine Survey,* 7:263–268.

Blokhuis, H.J., Jones, R.B., Geers, R., Miele, M. & Veissier, I. (2003) Measuring and monitoring animal welfare: transparency in the food product quality chain. *Animal Welfare* 12: 445–455.

Broom, D.M. & Johnson, K.G. (1993) *Stress and Animal Welfare.* Chapman & Hall, London.

Burchfield, J. (1975) *Lord Kelvin and the Age of the Earth.* Macmillan, London, pp. 93–109.

Burke, E. et al. (2006) Modelling the recent evolution of global drought and projections for the twenty-first century with the Hadley Centre Climate Model. *Journal of Hydrometeorology* 71:1113–1125. Cited in *The Independent* (UK) October 4, 2006.

CAST (Council for Agricultural Science and Technology) (1981) *Scientific Aspects of the Welfare of Food Animals.* Report no. 91.

Chiquitelli Neto, M., Paranhos da Costa, M.J.R. & Páscoa A.G. e Wolf, V. (2002) Manejo racional na vacinação de bovinos Nelore: uma avaliação preliminar da eficiência e qualidade do trabalho. In: Josahkian, L.A. (ed.), *5th Congresso das Raças Zebuínas.* ABCZ, Uberaba-MG, pp. 361–362.

Coelho, S.G. & Criação de Bezerros (2005) In: *II Simpósio Mineiro de Buiatria.* Belo Horizonte, 2005.

The Colorado Dairy Industry (2005) *Quick Facts Based on 2005 Production.*

Colles, F.M., Jones, T.A., McCarthy, N.D., Dawkins, M.S. & Maiden, M.C.J.M. *Campylobacter* infection of broiler chickens in a free-range environment. (submitted)

Darwin, C. (1872) *The Expression of the Emotions in Man and Animals.* Reprinted 1965 by the University of Chicago Press, London.

Dawkins, M.S. (1980) *Animal Suffering: The Science of Animal Welfare.* Chapman and Hall, London.

Dawkins, M.S., Edmond, A., Lord, A., Solomon, S. & Bain, M. (2004) Time course of changes in egg-shell quality, faecal corticosteroids and behaviour as welfare measures in laying hens. *Animal Welfare* 13:321–328.

Department for Environment, Food and Rural Affairs (2006) Animal Health 2006, The Report of the Chief Veterinary Officer. PB 12593. http://www.defra.gov.uk/animalh/cvo/report/2006/chap2.pdf. Accessed August 23, 2007.

Department for Environment, Food and Rural Affairs (2007a) Animal Health and Welfare Strategy indicators: livestock indicators: Farm assurance: core indicator 5.2. http://www.defra.gov.uk/animalh/ahws/eig/indicators/5-2.htm. Accessed August 23, 2007.

Department for Environment, Food and Rural Affairs (2007b) Animal Health and Welfare Strategy indicators: livestock indicators: Consumer response to different products: core indicator 5.1. http://www.defra.gov.uk/animalh/ahws/eig/indicators/5-1.htm. Accessed August 23, 2007.

Duncan, I.J.H. (1981) Animal rights – animal welfare: a scientist's assessment. *Poultry Science* 60:489–490.

Dunn, C.S. (1990) Stress reactions of cattle undergoing ritual slaughter using two methods of restraint. *Veterinary Record* 126:522–525.

Ehrenfeld, D. (2006) 'Friendly Fire.' *Resurgence* Nov/Dec No.239:23–25.

European Commission (2005) Special Eurobarometer: Attitudes of Consumers Towards the Welfare of Farmed Animals. http://europa.eu.int/comm/food/animal/welfare/euro_barometer25_en.pdf. Accessed May 6, 2006.

European Commission (2006) Community Action Plan on the Protection and Welfare of Animals 2006–2010. http://ec.europa.eu/food/animal/welfare/com_action_plan230106_en.pdf.

European Commission (2007) Special Eurobarometer: Attitudes of European Citizens towards Animal Welfare. http://ec.europa.eu/food/animals/welfare/survey/index-eu.pdf.

Ewbank, R., Parker, M.J. & Mason, C.W. (1992) Reactions of cattle to head restraint at stunning: a practical dilemma, *Animal Welfare* 1:55–63.

Farm Animal Welfare Council (1993) *Report on Priorities for Animal Welfare Research and Development.* PB 1310. FAWC, London.

Farm Animal Welfare Council (2005) *Report on the Welfare Implications of Farm Assurance Schemes.* FAWC, London.

Farm Animal Welfare Council (2006) *Report on Welfare Labelling.* FAWC, London.

Food and Agriculture Organization (2006a) In: Steinfeld, H. (ed.) *Livestock's Long Shadow. Environmental Issues and Options.* FAO, Rome. www.virtualcentre.org/en/library/key_pub/longshad/A0701E00.pdf.

Food and Agriculture Organisation (2006b) Press release: Livestock a major threat to environment. Remedies urgently needed. http://www.fao.org/newsroom/en/news/2006/1000448/index.html.

Fraser, A. (1972) *The Bull.* Osprey Publishing, Reading.

Fraser, A.F. & Broom, D.M. (1990) *Farm Animal Behaviour and Welfare*. CAB International, Wallingford.

Gomes, A., Madruga, C.R., Bianchin, I., Dode, M.A.N., Schenk, M.A.M., Honer, M.R., Pires, P.P., Kessler, R.H. & Silva, R.A. (1998) *Gado de corte: o produtor pergunta, a Embrapa responde*. Capitulo IV – Sanidade Animal, Campo Grande: Embrapa Gado de Corte, 500pp. www.cnpgc.embrapa.br/tecnologias

Grandin, T. (1980) Observations of cattle behavior applied to the design of cattle handling facilties. *Applied Animal Ethology* 6:19–31.

Grandin, T. (1982) Pig behavior studies applied to slaughter plant design. *Applied Animal Ethology* 9:141–151.

Grandin, T. (1990) Design of loading and holding pens. *Applied Animal Behavior Science* 28:187–201.

Grandin, T. (1992) Observations of cattle restraint devices for stunning and slaughtering, *Animal Welfare* 1:85–91.

Grandin, T. (1993) Handling facilities and restraint of range cattle. In: Grandin, T. (ed.) *Livestock Handling and Transport*. CAB International, Oxon, Wallingford. UK, pp. 75–94.

Grandin, T. (1996) Factors that impede animal movement in slaughter plants, *Journal of the American Veterinary Medical Association* 209:757–759.

Grandin, T. (1997a) The design and construction of facilities for handling cattle. *Livestock Production Science* 49:103–119.

Grandin, T. (1997b) *Survey of Stunning and Handling in Federally Inspected Beef, Veal, Pork, and Sheep Slaughter Plants*. USDA/ARS-3602-20-00. Project Number 3602-32000-002-08G. US Dept. of Agriculture, Beltsville, Maryland.

Grandin, T. (1997c) *Good Management Practices for Animal Handling and Stunning*. American Meat Institute Foundation, Washington, DC.

Grandin, T. (1998a) The feasibility of using vocalization scoring as an indication of poor welfare during cattle slaughter. *Applied Animal Behavior Science* 56:121–128.

Grandin, T. (1998b) Objective scoring of animal handling and stunning practices in slaughter plants. *Journal of the American Veterinary Medical Association* 212:36–39.

Grandin, T. (2000a) *Livestock Handling and Transport*, CAB International, Wallingford, Oxon, UK.

Grandin, T. (2000b) Effect of animal welfare audits of slaughter plants by a major fast food company on animal handling and stunning practices. *Journal of the American Veterinary Medical Association* 216:848–851.

Grandin, T. (2002) Cattle vocalizations are associated with handling and equipment problems at beef slaughter plants. *Applied Animal Behavior Science* 71:191–201.

Grandin, T. (2003) Transferring results of behavioral research to industry to improve animal welfare on the farm, ranch, and the slaughter plant. *Applied Animal Behavior Science* 81:215–228.

Grandin, T. (2005) Maintenance of good animal welfare standards in beef slaughter plants by use of auditing programs. *Journal of the American Veterinary Medical Association* 226:370–373.

Grandin, T. (2006) Progress and challenges in animal handling and slaughter in the US. *Applied Animal Behavior Science* 100:129–139.

THE FUTURE OF ANIMAL FARMING

Gregory, N.G., Wilkins, L.J., Eleperuma, S.D., Ballantyne, A.J. & Overfield, N.D. (1990) Broken bones in domestic fowls: effect of husbandry system and stunning method in end-of-lay hens. *British Poultry Science* 31:59–69.

Hall, G.A. (1987) Comparative pathology of infection by novel diarrhea viruses. In: *Ciba Foundation Symposium, 128. Enteropathology of Diarrhea Viruses.* Ciba Foundation, Wiley, Chichester, pp. 192–217.

Hall, J.V., Brajer V. & Lurmann F.W. (2006) *The Health and Related Economic Benefits of Attaining Healthful Air in the San Joaquin Valley.* Institute for Economic and Environmental Studies, California State University, Fullerton.

Harrison, R. (1964) *Animal Machines. The New Factory Farming Industry.* Ballantine Books, New York.

Henson, R. (2006) *The Rough Guide to Climate Change.* Rough Guides Ltd, London, UK.

Hillman, M. (2004) *How We Can Save the Planet.* Penguin, London, UK.

Houghton, J. (2004) *Global Warming, The Complete Briefing,* 3rd edn. Cambridge University Press, Cambridge, UK.

HRH The Prince of Wales (2007) A sense of harmony. *Resurgence* May/June No. 242: 14–16.

Humphrey, T. (2006) Are happy chickens safer chickens? Poultry welfare and disease susceptibility. *British Poultry Science* 47:379–391.

Intergovernmental Panel on Climate Change (2007) 4th Assessment Report. http://www.ipcc.ch. Accessed July 24, 2007.

Jarvis, S., D'Eath, R.D. & Fujita, K. (2005) Consistency of piglet crushing by sows. *Animal Welfare* 14: 43–51.

Jarvis, S., Reed, B.T., Lawrence, A.B., Calvert, S.K. & Stevenson, J. (2004) Perinatal environmental effects on maternal behaviour, pitutitary–adrenal activities and the progress of parturition in the primiparous sow. *Animal Welfare* 13:171–181.

Jones, T., Feber, R., Hemery, G., Cook, P., James, K., Lamberth, C. & Dawkins, M. (2007) Welfare and environmental benefits of integrating commercially viable free-range broiler chickens into newly planted woodland: a UK case study. *Agricultural Systems* 94:177–188.

Kilgour, R. (1971) Animal handling in meat works pertinent behavior studies. In: *Proceedings of the 13th Meat Industry Research Conference, Hamilton, New Zealand,* pp. 9–12.

Kristal, M.B. (1980) Placentophagia: a behavioral enigma (or De gustibus non disputadum est). *Neurosciencs and Biobehavoral Reviews* 4:141–150.

Lash, J. (2005) quoted in Amos, J. (2005) 'Study highlights global decline' BBC News. http://news.bbc.co.uk/1/hi/sci/tech/4391835. stm. Accessed July 24, 2007.

Lidfors, L. (1994) Mother-young Behaviour In Cattle. Doctoral Thesis, University of Agricultural Sciences, Swedish, Report 33.

Locke, J. (1690) Essay *Concerning Human Understanding.* Book II, 9.

Macfarlane, R. (2006) Turning points. In: *Burning Ice, Art and Climate Change.* Cape Farewell, London, UK.

Machado Neto, R., Faroni, C.E., Pauletti, P. & Bessi, R. (2004) Levantamento do manejo de bovinos leiteiros recém-nascidos: desempenho e aquisição de proteção passiva. *Revista Brasileira de Zootecnia* 33:2323–2329.

Main, D.C.J. & Green L.E. (2000) A descriptive analysis of the operation of the Farm Assured British Pigs. *Veterinary Record* 147:162–163.

Main, D.C.J., Whay, H.R., Green, L.E. & Webster, A.J.F. (2003) Effect of the RSPCA Freedom Food scheme on dairy cattle welfare *Veterinary Record* 153:227–231.

Marris, E. (2006) Grey matters. Many scientists have nuanced views on animal research, but they are rarely heard. *Nature* 444:808–810.

Mason, G. & Mendl, M. (1993) Why is there no simple way of measuring animal welfare? *Animal Welfare* 2:301–320.

Mason, G., Cooper, J. & Clareborough, C. (2001) Frustrations of fur farmed mink. *Nature* 410: 35–36.

McCarthy, M. (2007) Final Warning. *The Independent* February 3, 2007. London, UK.

Mayfield, Bennett & Tanter (2007) United Kingdom report on focus group research. In: Evans, A. & Miele, M. (2007) *Consumers' Views About Farm Animal Welfare Part 1: National Reports Based on Focus Group Research.* Quality reports no. 4. pp. 115–155.

Midgley, M. (1983) *Animals and Why They Matter.* Penguin, Middlesex, UK.

Millennium Ecosystem Assessment (2005a) http://www.millenniumassessment.org/en/index.aspx. Accessed July 24, 2007.

Millennium Ecosystem Assessment (2005b) Quoted In Radford, T. (2005) 'Two-thirds of world's resources 'used up'.' http://www.guardian.co.uk/international/story/0,3604,1447863,00.html. Accessed July 24, 2007.

Monbiot, G. (2006) *Heat: How to Stop the Planet Burning* Penguin, London, UK.

Murphy, D. (2006) The vocal point: Media darling's second act plays to packed house. www.cattlenetwork.com/content/asp?contenid=83107. Accessed November 10, 2006.

National Farmers Union. (1995) Animal Welfare. Information leaflet. http://www.virtual-centre.org/en/library/key_pub/longshad/A0701E00.htm

Nordlund, K., Cook N.B., & Octzel, G.R. (2004) Investigation strategies for laminitis problem herds. *Journal of Dairy Science* 87(E Suppl): E27–E35.

Olsson, I.A.S., Keeling, L.J. & McAdie, T.M. (2002) The push-door for measuring motivation in hens: an adaptation and a critical discussion. *Animal Welfare* 11:1–10.

Paranhos da Costa, M.J.R., Zuin, L.F.S. & Piovesan, U. (1998) Avaliação preliminar do manejo pré-abate de bovinos no programa de qualidade da carne bovina do Fundepec. *Relatório Técnico*, 21pp.

Paranhos da Costa, M.J.R. (2004) Comportamento e bem-estar de bovinos e suas relações com a produção de qualidade. In: *41a Reunião da Sociedade Brasileira de Zootecnia, 2004, Campo Grande-MS.* Sociedade Brasileira de Zootecnia/Embrapa Gado de Corte, Campo Grande-MS, v. 41, pp. 260–268.

Paranhos da Costa, M.J.R., Schmidek, A. & Toledo, L.M. (2006a) *Boas práticas de manejo: bezerros ao nascimento.* Editora Funep, Jaboticabal, 36pp. e-book: www.grupoetco.org.br

Paranhos da Costa, M.J.R., Toledo, L.M. & Schmidek, A. (2006b) *Boas práticas de manejo: vacinação.* Editora Funep, Jaboticabal, 32pp. e-book: www.grupoetco.org.br

Peterson, D. (2006) *Jane Goodall, the Woman who Redefined Man.* Houghton Mifflin, Boston, Mass.

Phillips, H. (2006) Known unknowns. *New Scientis* December 16, p. 28.

Rawles, K. (1997) Conservation and animal welfare. In: Chappell, T.D. (ed.) *The Philosophy of the Environment.* Edinburgh University Press, Edinburgh, UK.

Rawles, K. (2006) Sustainable development and animal welfare: the neglected dimension. In: Turner, J. & D'Silva, J. (eds) *Animals, Ethics and Trade.* Earthscan, London, UK.

Rollin, B.E. (1981) *Animal Rights and Human Morality.* Prometheus, Buffalo, New York.

Rollin, B. (1992) *Animal Rights and Human Morality*, revised edition. Prometheus, Buffalo, New York.

Rollin, B.E. (1995) *Farm Animal Welfare.* Iowa State University Press, Ames, Iowa.

Rollin, B.E. (2006) *Animal Rights and Human Morality* 3rd edn. Prometheus, Buffalo, NY.

Rowlatt, J. (2007) Meet daisy the cow – global climate's enemy number one. http://www.bbc.co.uk/blogs/newsnight/2007/02/meet_daisy_the_cow_global_climates_enemy_number_on.html. Accessed July 24, 2007.

Royal Society for Prevention of Cruelty to Animals (2001) *RSPCA Welfare Standards for Dairy Cattle.* RSPCA, Horsham, UK.

Rushen, J. (1991) Problems associated with the interpretation of physiological data in the assessment of animal welfare. *Applied Animal Behavior Science* 28:381–386.

Savory, C.J. (2004) Laying hen welfare standards: a classic case of 'power to the people.' *Animal Welfare* 13:S153–S158.

Schmidek, A., Paranhos da Costa, M.J.R., Mercadante, M.Z. & Toledo, L.M. (2006) The effect of newborn calves vigour in their mortality probability. In: *40th International Congress of the International Society for Applied Ethology, 2006, Bristol, UK.* pp. 257.

Scott, K., Taylor, L., Gill, B.P. & Edwards, S.A. (2006) Influence of different types of environmental enrichment on the behaviour of finishing pigs housed in two different systems. ! Hanging toy v. rootable substrate. *Applied Animal Behaviour Science* 99:222–229.

Serpell, J. (1986) *In the Company of Animals.* Blackwell, Oxford.

Silva, L.C.M., Madureira, A.P., Ferreira, P.C.M., Lazaro, L.O., Simões, M., Abud, G.C. & Paranhos da Costa, M.J.R. (2006) A adoção do manejo racional melhora o bem-estar de bezerros leiteiros. In: *I Congresso Internacional de Conceitos em Bem-Estar Animal, 2006.* Rio de Janeiro. WSPA, Rio de Janeiro.

Singer, P. (1991) *Animal Liberation*, 2nd edn. Thorsons, London, UK.

Singer, P & Mason, J. (2006) *Eating: What We Eat and Why it Matters.* Arrow Books, Random House, London. First published 2006 in the US under the title *The Way We Eat: Why Our Food Choices Matter* by Rodale Inc.

Singer, P. & Regan T. (1976) *Animal Rights and Human Obligations.* Prentice-Hall, New Jersey.

Steinfeld, H., Wassenaar, T. & Jutzi, S. (2006) Livestock production systems in developing countries: status, drivers, trends. *Revue Scientifique et Technique-Office International des Epizooties* 25(2):505–516.

Stern, Sir N. (2006) *Stern Review on the Economics of Climate Change.* H.M.Treasury, London, UK.

Stolba, A. & Wood-Gush, D.G.M. (1984) The identification of behavioural key features and their incorporation into a housing design for pigs. *Annales de Recherches Vétérinaires* 15:287–298.

Stolba, A. & Wood-Gush, D.G.M. (1989) The behaviour of pigs in a semi-natural environment. *Animal Production* 48:419–425.

Tanida, H.L., Miura, A., Tanaka, T. & Yoshimoto, T. (1996) Behavioral responses of piglets to darkness and shadows. *Applied Animal Behavior Science* 49:173–813.

Thomas, C.D., Cameron, A. & Green, R.E. et al (2004) Extinction risk from climate change. *Nature* 427:145–148.

Toates F. (1995) *Stress. Conceptual and Biological Aspects.* Wiley, New York.

Toledo, L.M. (2005) Fatores intervenientes no comportamento de vacas e bezerros do parto até a primeira mamada. Tese de Doutorado, Faculdade de Ciências Agrárias e Veterinárias – UNESP, 73pp.

Tseimazides, S.P, Barbalho, P.C, Ciocca, J.R.P, Paranhos da Costa, M.J.R.(2005) Avaliação de efeitos de um programa de treinamento no manejo de bovinos no desembarque e no pH das carcaças. In: *42a Reunião Anual da Sociedade Brasileira de Zootecnia, 2005, Goiânia-GO. A produção animal e o foco no agronegócio.* Sociedade Brasileira de Zootecnia, Goiânia-GO, v. 42. pp. 1–3.

Tudge, C. (2004). *So Shall We Reap. What's Gone Wrong with the World's Food – and How to Fix it.* Penguin Books, London.

Tudge, C. (2007) *Feeding People is Easy.* Pari Publishing, Pari, Italy.

USDA/NASS (2005) *Livestock.* United States Department of Agriculture/National Agricultural Statistical Service.www.usda.gov/nass/pubs/agstats.htm

USDA/NASS (2006) *Milk Production and Milk Cows.* United States Department of Agriculture/ National Agricultural Statistical Service.www.usda.gov/nass/pubs/agstats.htm

Van Putten, G. & Elshoff, G. (1978) Observations on the effect of transport on the well being and lean quality of slaughter pigs. *Animal Reguation Studies* 1:247–271.

Vansickle, J. (2002) Profits slow decline in hog farm numbers. *Natural Hog Farmer* May 15.

Vessier, I. & Evans (2007) Rationale behind the Welfare Quality® assessment of welfare. In: Veissier, I., Forkman, B. & Jones, B. (eds), *Proceeding of Second Welfare Quality® Stakeholder Conference, 3–4 May 2007, Berlin Germany. Assuring Animal Welfare: From Societal Concerns to Implementation,* pp. 19–22. http://www.welfarequality.net/publicfiles/ 36059_25646376170_200705090907523_2244_Proceedings_2nd_WO_Stakeholder_ conference_3_4_May_2007.pdf

Voisinet, B.D., Grandin, T., O'Connor, S.F., Tatum, J.D. & Deesing, M.J. (1997) *Bos indicus*-cross feedlot cattle with excitable temperaments have tougher meat and a high incidence of borderline dark cutters. *Meat Science* 46(4):367–377.

Waal, de F. (2006) *Primates and Philosophers: How Morality Evolved.* Princeton University Press, Princeton.

Warriss, P.D., Brown, S.N. & Adams, S.J.M. (1994) Relationship between subjective and objective assessment of stress at slaughter and meat quality in pigs. *Meat Science* 38:329–340.

Weary, D.M., Braithwaite, L.A. & Fraser, D. (1998) Vocal responses to pain in piglets. *Applied Animal Behavior Science* 56:161–172.

Webster, J. (1994) *Animal Welfare: A Cool Eye Towards Eden.* Blackwell Science, Oxford.

Whay, H.R., Leeb, C., Main, D.C.J., Green, L.E. & Webster, A.J.F. (2007a) Preliminary assessment of finishing pig welfare using animal-based measurements *Animal Welfare* 16, 209–211.

Whay, H.R., Main, D.C.J., Green, L.E., Heaven, G., Howell, H., Morgan, M., Pearson, A. & Webster A.J.F. (2007b) Assessment of the behaviour and welfare of laying hens on free-range units. *Veterinary Record* 161:119–128.

White, R.G., DeShaver, J.A., Treassler, C.J., Borcher, G.M., Davey, S., Warninge, A., Parkhurst, A.M., Milanuk, M.J. & Clems, E.T. (1995) Vocalization and physiological response of pigs during castration with and without anesthetic. *Journal of Animal Science* 73: 381–386.

World Commission of Environment and Development (1987) *From One Earth to One World: An Overview.* Oxford University Press, Oxford.

World Wildlife Fund (2004) *Living Planet Report.* http://www.eldis.org/go/display/?id=17012 &type=Document. Accessed July 24, 2007.

Index